1-5-76

The
Political Economy
of North Sea Oil

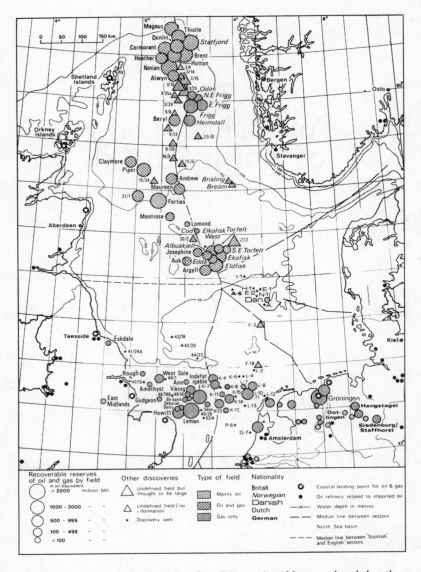

Frontispiece. North Sea Oil and Gas Discoveries. Map produced by the Economisch Geografisch Institute, Rotterdam, but English/Scottish median line added by the authors.

The
Political Economy
of North Sea Oil

D. I. MacKay
& G. A. Mackay

WESTVIEW PRESS
BOULDER, COLORADO

ISBN: 0–89158–515–X

Published in 1975 in London, England
by Martin Robertson and Company Ltd.

Copyright 1975 in London, England
by Martin Robertson and Company, Ltd.

Published in 1975 in the United States of America
by Westview Press, Inc.
  1898 Flatiron Court
  Boulder, Colorado 80301
  Frederick A. Praeger, Publisher and Editorial Director

Library of Congress Cataloging in Publication Data

MacKay, Donald Iain
    The political economy of North Sea oil.
    Bibliography: p.
    Includes index.
    1.   Petroleum industry and trade—Great Britain.
2.   Great Britain—Economic conditions—1945—
I.   Mackay, George A., joint author.   II.   Title.
HD9571.5.M3   338.2′7′2820941   75–25633
ISBN 0–89158–515–X

Printed and bound in Great Britain

1892042

Contents

Tables

Preface

The oil and gas discoveries in the North Sea offer the prospect of substantial economic advantages to the UK – and also the prospect of major political strains and tensions. The one follows on from the other. The economic significance of North Sea oil lies in the fact that, even in the hostile environment of the North Sea, the resource costs of winning the oil and gas are small relative to their market value. In recent weeks we have heard much about the high and rising cost of North Sea production. That costs are high by world standards is indisputable, but so, too, is the market price for oil. In this book we have based our calculations on estimates which, in our opinion, take full account of past increases through 1974 and, indeed, our cost figures are significantly higher than those put forward by other commentators. Nonetheless, it remains true that the resource cost is low relative to the market value of production. In consequence, the major economic benefits will accrue through the balance of payments to the Central Government in the form of royalties and taxation revenue. The reasons for this outcome, and the likely economic benefits for the UK, are explored in chapters 1–5 which investigate, in turn, the importance of oil in the UK economy; government policy toward exploration, production and taxation; the extent of North Sea oil and gas reserves and likely production; the nature of the industrial activities in the North Sea and on land; and the implications of indigenous oil and gas production for energy requirements and supply, the UK balance of payments and government revenue.

Chapters 6–8 examine the economic benefits accruing to Scotland, off whose shores the major discoveries have been made. We look at the extent of employment creation in North Sea activities, the development of government policy on land and the economic consequences which would flow from different constitutional and political arrangements between Scotland and the remainder of the UK. We will argue that the direct benefits accruing to Scotland from present arrangements *must* be small relative to the benefits accruing to the Central Government and that the former are likely to be insufficient to eradicate the low incomes and high unemployment traditionally experienced in Scotland. Such a situation might be sustainable in a nation of regions, but our view of political realities is based on the supposition that the UK is best regarded, historically, socially and culturally, as a union of nations. This does not mean that the constituent nations should each pursue its own economic self-interest to the detriment of other considerations, but it does mean that they are unlikely to be oblivious to that self-interest. To put it bluntly, present policies offer the Scots altruism and relative poverty. Unless some viable intermediate position can be found, they are likely to prefer avarice and relative affluence.

In writing this book we have been greatly assisted by our friends and colleagues in the oil industry, in other commercial activities and in academic life. Alan Peters, George Innes and other individuals in Shell International and Shell U.K. commented on some of our early drafts and we were also assisted by employees of British Petroleum, Occidental and a number of other oil companies. Tony MacIntosh of Wood, MacKenzie provided invaluable information on capital and operating costs in the North Sea and, together with Ian Smith and Frank Malcolm of Bell, Lawrie, Robertson, read and criticised all our initial drafts. Peter Odell supplied Figure 3.1 and our colleagues in the Department of Political Economy, Aberdeen University – Max Gaskin, Joe Kemp, Niall Trimble and Anne Moir – provided criticism, advice and encouragement, not necessarily in that order. Alison Crockett provided an unfailing and unflagging source of technical help and Anne Bain and Willa Fraser typed the numerous drafts with more patience and good humour than we deserved. We add the usual rider that we alone are responsible for the views expressed and for any errors of commission and omission.

Our manuscript was completed in early November 1974. There have been a number of new developments since that date, but they appear to confirm, rather than contradict, the basic assumptions underlying our major arguments. Thus, our estimate of the likely build up of oil production was made on the implicit assumption that there would be no government attempt to regulate output before 1980. This position was officially endorsed by the Secretary of State for Energy in his Commons statement of December 6, 1974 and that statement provides further support for our view that oil refining and processing will provide relatively few employment opportunities in Scotland. Recent announcements regarding taxation policy underline our argument for a flexible regime which is generous to marginal fields. Only one minor amendment appears necessary. Production from the Argyll Field will be slightly delayed on the timing shown in Appendix 2.1, and this must have some adverse impact on the field's profitability. However, as the field is a small one the amendment is not of major significance.

D. I. MacKay
G. A. Mackay

King's College,
University of Aberdeen.
January 1975.

TO ROSS
and
TO CROMARTY

"Beware of false prophets, which come to you in sheep's clothing, but inwardly they are ravening wolves."

Matthew 7, v. 15

CHAPTER 1

Oil and the UK Economy

There are a number of reasons for expecting [oil prices] not to increase ... Competition is strong both between companies and between sources of supply, and the surplus of crude oil seems likely to persist for many years despite the expansion in world demand. (Fuel Policy, Cmnd 3438, 1967, para. 26)

1.1 INTRODUCTION

Looking back across the short period in which the UK, along with other oil-importing countries, has suffered a rapid and unprecedented rise in the price of its imported crude oil, and reductions in supplies imposed by producer countries and by two Arab–Israeli wars, this quotation emphasises the frailty of human judgement. Forecasting energy requirements and possibilities seems to have advanced little since 1866 when Gladstone, advised by that eminent economist Jevons, forecast a British demand for coal of 3,000 million tons by 1970.[1]

In 1967 there appeared to be little reason to question the central tenet of British fuel policy. Over the 1960s world demand for primary energy had risen by some 5 per cent per year. This had been accommodated with some ease by a cheap and plentiful supply of oil. On this basis was erected the spectacular economic growth achieved by most industrialised economies in the 1960s. Without such an apparently inexhaustible supply of cheap energy, economic growth would have been subject to more severe restraints imposed by the increasing costs of coal production and the disappointing progress towards cheap nuclear power.

By 1970, 60 per cent of European primary energy requirements were met by imports. Japan's dependence on imports was almost total and there was a clear prospect of the huge American market emerging as a substantial net importer of energy in the 1970s. It

1

was the very success of oil in sustaining economic growth and in penetrating new markets which exposed the major industrial powers to increased dependence on a small group of oil-exporting countries. In the 1970s the relationships between producer countries on the one hand and the international oil companies and consumer countries on the other have been transformed. From a period of an incipient oil surplus we have emerged suddenly into a world of threatened shortage, where the major oil producers, provided they can continue to act together, wield substantial economic and political power.

The true significance of the oil and gas discoveries in the UK sector of the North Sea can only be appreciated against this background. This chapter will look at the place of oil in the UK economy and at how our position compares with that of other industrialised countries; at the increasing dependence of the major industrial powers on oil imports and the factors which dictated this increased reliance; at the emergence of the Middle East producers and the Organisation of Petroleum Exporting Countries (OPEC) which has revolutionised oil markets and oil prices; and at the probable future trend of oil prices and the likely effects on the UK economy.

1.2   THE DEMAND FOR PRIMARY ENERGY

In the 1960s UK oil consumption increased by almost 8 per cent per annum as cheap oil captured an increasing share of coal's traditional market. The pattern of oil consumption remained remarkably constant as each major segment of oil demand increased at much the same rate. In both Britain and OECD Europe fuel oil is the major component of oil demand, accounting for 40–45 per cent of total consumption, compared with the 13–15 per cent of petrol. The pattern of oil consumption is very different in the USA, where fuel oil is only 15 per cent and petrol 40 per cent of oil consumption. The difference springs from the greater share of primary fuel needs met in the USA by coal and natural gas – 54 per cent compared with 30 per cent in OECD Europe, although in the UK these fuels account for 52 per cent of energy requirements[2] – and from the high density of US car ownership combined with the low miles-per-gallon performance of US cars. The large demand for petrol makes the United States potentially an extremely attractive market for North

Sea oil, which is a high-quality oil with a low sulphur content and a high yield of lighter fractions, e.g. gasoline, naphtha (which may be used *inter alia* as a petro-chemical feedstock) and middle distillates. The heavy-fuel yield will be low and it is probable that much of North Sea oil production will be exported,[3] particularly to northern Europe but also to the American market, while the UK continues to import heavy crudes from Middle East producers to satisfy its large requirement for fuel oil.

While there is no obvious alternative to oil for road passenger transport and air transport, there is much greater possibility of substitution between oil and other fuels as suppliers of other types of primary energy. In 1950 coal was still king, accounting for 90 per cent of primary energy requirements. Oil's share of the market increased only slowly until 1957; but from that date its advance has been rapid despite substantial discrimination in favour of coal, the prime indigenous source of fuel. Consumers, given an unrestricted choice, would have chosen a greater reliance on cheaper and more convenient oil than on dearer and less convenient coal. However, UK energy policy was built on the view that this would have given insufficient weight to wider considerations.[4] In brief, a greater reliance on oil would have involved heavier balance of payments costs, greater risk because of the increased dependence on external supplies from a potentially unstable region of the world and a quicker rundown of coal, with the attendant economic, social and political costs of heavier unemployment in mining areas.

Thus, while there was a steady move to greater dependence on oil imports, successive governments intervened to protect the British coal industry from the full effects of external competition. In 1961 an additional tax was placed on heavy-fuel oil, initially to raise revenue, but continued to protect coal. By 1967 the tax amounted to 40 per cent of the pre-tax price of fuel oil to industrial users, a level of protection for coal much greater than that afforded to almost any other sector of the economy. The electricity and gas industries were forced to use more coal than they would have chosen and the indigenous industry was further protected by the restrictions placed on coal imports from the Soviet bloc and the USA. It was always recognised that coal could not maintain its former virtual monopoly of the fuel market and that protection simply slowed and eased the transition to a two-fuel economy. Indeed, despite substantial increases in productivity, the labour-intensive coal industry found it increasingly difficult to compete with oil. The pre-customs duty price of oil fell over the 1960s because the economies made by the major oil companies in downstream operations (bulk tankers, improved refining techniques) and increasing competition offset the higher taxes levied by producer countries.

With the discovery of natural gas in the southern North Sea in the mid-1960s and the hoped for developments in nuclear power, it was coal and not oil that had to give ground. The central assumption of the 1967 White Paper was 'that regular and competitively priced supplies of oil will continue to be available . . . as they have been in increasing quantities over the past years'. The supply of oil was considered perfectly elastic at the going price so that the new fuels would squeeze coal's share of the energy market. Whereas in 1966 there was a two-fuel economy, coal 60 per cent and oil 40 per cent, by 1975 a four-fuel economy was envisaged with the market divided between coal (34 per cent), oil (41 per cent), natural gas (14 per cent) and nuclear power (10 per cent). In the event, oil had captured almost half of the total market by 1972 and an even greater squeeze on coal was avoided only because of the failure to make the expected progress towards developing cheaper electricity from the new generation of nuclear power stations.

The chief criticism directed against UK fuel policy in the 1960s was that continued central intervention seriously distorted the market and increased the basic price of energy. The critics would have preferred a more rapid contraction of the coal industry by exposing it to greater competition. In the circumstances of the 1960s this would have meant greater dependence on oil. As it is, while the UK has become much more dependent on oil imports through time, it is still noticeably less dependent on oil imports than most other industrialised countries. The weight attached to security of supply in the 1960s may pay dividends in the 1970s. With the exception of Luxembourg, coal supplies a higher proportion of primary energy needs in the United Kingdom than in *any* of the nineteen countries in OECD Europe.

The UK is, then, less exposed than the industrialised countries of Europe to the economic dangers inherent in the advent of a seller's market in oil. However, the invalid is not always consoled by the thought that others face greater suffering, and the UK's ability to minimise the effects of increased oil prices does depend, in the short run, on the greater size of the indigenous coal industry, a fact that has not escaped the attention of miners! Although the UK's relative dependence on oil may be less than that of others, its absolute dependence is still very substantial. In the immediate future relatively little can be done to avoid the consequences of the steep rise in oil prices of the 1970s and even in the medium term the UK's fortunes depend on the extraordinary luck of finding major oil and gas deposits in the North Sea.

The extent of this dependence and its effects on the balance of payments even before the price rises of 1973–74 are illustrated by table 1.1.

TABLE 1.1 UK NET IMPORTS OF OIL: VOLUME AND VALUE

| | Crude | | Refined Products | Net* Deficit |
	(million tons)	(value per ton)	(million tons)	(£ million)
1950	9.2	8.03	8.7	181.1
1955	27.9	8.04	2.2	256.7
1960	44.6	7.44	4.7	375.3
1965	65.4	6.47	9.6	495.4
1970	100.0	7.02	3.7	754.1
1972	102.3	8.98	4.7	953.0

*Imports (c.i.f.) less exports (f.o.b.)

Source: Department of Trade and Industry *United Kingdom Energy Statistics* (1973) tables 122 and 126

The volume of net imports of refined products tended to fall from 1950 because, with government encouragement aimed at saving foreign exchange, refining capacity was expanded in line with increased home consumption needs. Crude oil imports increased tenfold in the two decades from 1950, but by 1970 the net deficit, as measured in table 1.1, was only four times the 1950 level. The price of crude had therefore declined in money terms and, allowing for inflation, even more in real terms over 1950–70. It was this experience that explains the increased share of oil in the primary energy market and the optimism of the 1967 White Paper. Indeed, the balance of payments cost of oil imports was less than table 1.1 implies. The data presented are the only consistent series from which one can judge the *trend* of the balance of payments cost of oil imports, but as imports are valued c.i.f. and exports f.o.b. the *level* of costs is overstated. For this, imports should be measured f.o.b. (to reflect the use of British tankers, investment in foreign oil companies in the UK, cost of headquarters and other services of British oil companies, together with profits remitted to the UK), and this would yield a 'true' net cost only some two-thirds of that shown in table 1.1. On this basis, the net deficit in 1972 was £657m.

The increasing volume of oil meant an increasing balance of payments cost, but although efforts were made to contain these costs, for example through encouraging additional investment in refining capacity in the UK, the burden imposed on the current account of the balance of payments through increased dependence on oil imports was cushioned, to a considerable extent, by the fall in the real price of oil imports. In the 1960s there was little reason to suppose that this position would end, but it did. From 1970 the price of crude oil began to increase. Over 1970–72 the volume of imported crude rose very little, but the net deficit rose significantly, as table 1.1 shows. The era of cheap oil had come to an end, at just the period when oil had emerged as Britain's major source of energy.

1.3   THE RISE AND RISE OF OPEC

World energy requirements rose at an annual rate of almost 5 per cent over 1960–70 and oil's share of primary fuels expanded from 33 per cent to 44 per cent over the same period. Crude oil production was 1019 million tons in 1959 and 2353 million tons in 1970. This massive increase, to meet rising demand, was accomplished without any increase in the price of oil. On the contrary, the real price of oil fell through most of the 1960s, as a result of an apparently inexhaustible supply of Middle East crude and competition from new sources such as Libya and Nigeria.[5] The 1960s were a decade in which an incipient surplus of oil supplies at ruling prices always threatened to thwart the attempts of producing countries to increase their revenue through levying higher taxes. Although the host countries were able to increase their 'take', this was at the expense of oil company profits. The market price of oil drifted downward in real terms and the cheap and cheapening supply of oil fuelled the growth of world energy requirements.

These developments were made possible by rapid expansion of Middle East oil production from 257 million tons in 1960 to 689 million tons in 1970. This area emerged to take a dominant position in world production and, more significantly, in world trade in oil and in proved world reserves (see figure 1.1).

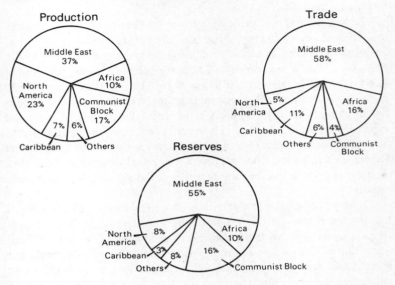

Figure 1.1   Share of world oil production, trade and proved reserves, 1973 (from BP *Statistical Review of the World Oil Industry* (1973))

The economic and strategic power of the Middle East producers derives from their dominant position in world trade in oil and from the huge size of their crude oil reserves. Only the increase in African production, from Libya, Algeria and Nigeria, has prevented the European consuming nations being completely dependent on Middle East production, but Middle East countries are still the dominant suppliers. In 1970 Middle East producers, together with Libya, supplied 76 and 84 per cent of the oil imports of OECD Europe and the UK respectively, and 86 per cent of Japan's needs.[6]

The vast majority of the advanced industrial nations are, then, substantial oil consumers with no significant indigenous oil production or reserves, and they depend for their imports on a handful of small states concentrated in one of the most strategically vulnerable areas of the world. Seven of the twelve producing countries that constitute OPEC – Saudi Arabia, Kuwait, Iran, Iraq, Libya, Qatar and the United Arab Emirates – with a combined population of only 60 million control almost 70 per cent of world trade in oil. As we have seen, the consuming countries enjoyed a buyer's market in the 1960s. What we now have to consider is why the situation has been so transformed in the 1970s.

The estimated f.o.b. price of Persian crude was $1.75 per barrel[7] in 1949 and prices changed little until they were raised in 1953, and again in 1957 to reach a peak which was not regained until the 1970s. By 1960 the price was down to $1.50 per barrel. This slipped further to $1.17 in 1965 and was still less than $1.25 per barrel in 1969.[8] In an inflationary world it may appear remarkable that the oil prices fell in money terms, particularly if we remind ourselves of the rapid expansion of world demand. However, price changes depend on the interaction of demand and supply conditions, and when supply side factors are considered what appears to be remarkable is not the decline in the real price of oil, but the fact that it did not decline much more rapidly.

The most careful study of crude oil prices and supply conditions has estimated that in the 1960s in the *most expensive* area of production in the Persian Gulf, the cost of producing oil at the wellhead (including a 20 per cent return on investment) was only 10 cents per barrel.[9] Nor was there any indication that the long-run marginal cost curve was steeply upward sloping. On the contrary, 'for at least 15 years we can count on, and must learn to live with, an abundance of oil that can be brought forth from fields operated in the Persian Gulf at something between 10 and 20 cents per barrel at 1968 prices'.[10] In short, even in the era of cheap oil in the 1960s there was a huge gap between production cost and the market price of the product. To understand how this continues, and how the gap between market prices and the minimum production cost has now become an awesome chasm, requires some investigation of the market structure of the industry and relationships between host countries, oil companies and consumer nations.

The international oil industry has traditionally been a highly concentrated oligopolistic structure. It was dominated by the eight 'majors', five of which are American, one Anglo-Dutch, one British and one French. Each of these companies – Standard Oil of New Jersey (now Exxon), Texaco, Gulf, Standard Oil Company of California, Mobil, Royal Dutch-Shell, British Petroleum and compagnie Francaise des Pétroles – has shown a high degree of vertical integration through exploration, production, transport and refining to final distribution. The position of the majors came under increasing challenge from the so-called 'independent' companies in the late 1950s and 1960s, but it remains strong. The interdependence of oligopolistic producers has given rise to a classic textbook situation where each individual firm (and each host country) believes that price cuts will be followed by its competitors so that all face a fall in revenue. Competition therefore has often taken a non-price form,[11] and this, together with the marketing power and efficiency of the small group of majors, is the first important factor underpinning a market price which is significantly above the long-run price of oil production.[12]

Historically the oil majors have operated via concession agreements with host countries which have given them exploration and production rights in return for royalty payments and a division of profits between the company and the host country. Given the vertical integration of the majors, and their interest in arranging pricing policies to minimise total tax liabilities over all transactions, the host country must have some means of applying effective fiscal control. This was provided by posted prices, which since 1960 have been a tax reference price quite unrelated to the prices that might be established by market forces.

This was not always so. In 1957 posted prices had been raised with favourable effects on oil company profits. The independents, who had by then acquired significant production, were unable to penetrate the US market owing to the imposition of import quotas and turned instead to the rapidly growing European market for energy. They could establish a foothold only through price competition; and the posted price and freight rate system, which had established the 'market' price of delivered oil, came under severe pressure. The independents' share of oil production rose between 1955 and 1960 from 8 to 16 per cent, of refining from 19 to 26 per cent, and by the latter date they controlled 30 per cent of product sales. The majors reacted by revising posted prices downwards in 1959, and again in 1960, which had the direct effect of significantly reducing the revenue per barrel of the producing countries. So the Organisation of Petroleum Exporting Countries (OPEC) was born with a view to promoting concerted action on the part of the major exporting nations.

The history of OPEC, and the running battle in the 1960s between the host countries and the oil majors, has been dealt with at length elsewhere[13] and need not be repeated in detail. Suffice to say that the oil majors were in retreat through most of the period. Their strength in production and marketing was weakened by the increased importance of the independents. The return on their capital was squeezed between an increased government take and the stickiness of final prices, and their role as producers came under serious threat from host governments. OPEC successfully raised the host government's take, and in the buyers' market with falling realised prices this was accomplished at the expense of company profits. Between 1960 and 1970 the host government's take increased from 70.8 to 86.0 cents per barrel, while the average return from the oil majors declined from 56.5 to 32.7 cents, equivalent to a rate of return on capital invested of 14 per cent and 11 per cent.[14]

This trickle of concessions became a flood in the 1970s. Libya was the first country to secure an increase in posted prices in 1970, but this was quickly followed by the Teheran and Tripoli agreement in 1971, the Geneva agreement of 1972 and the Teheran escalation of 1973. As a consequence of these agreements the combined revenue of OPEC countries rose from some $4500m in 1966 to $7500m in 1970, $12,000m in 1971 and $15,000m in 1972,[15] and was estimated to increase, given the arrangements existing in late 1972, to $56,000m in 1980. The Arab-Israeli war of 1973 and the subsequent action of the OPEC countries make even these estimates seem strangely conservative. The Kuwait announcement in October 1973 by OAPEC (The Organisation of Arab Petroleum Exporting Countries) threatened a successive 5 per cent monthly reduction of output 'until Israeli withdrawal is completed from the whole Arab territories occupied in June 1967 and the legal rights of the Palestinian people are restored'. In

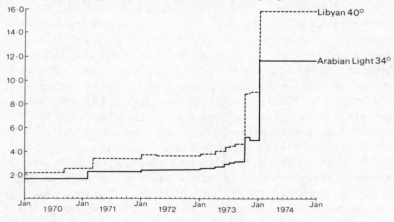

Figure 1.2   Crude oil posted prices (posting at f.o.b. loading port), 1970–4, Libyan 40° and Arabian Light 34°

addition, a unilateral decision to raise posted prices by 70 per cent was announced and, in January 1974, a doubling of posted prices produced a further twist of the screw. Figure 1.2 shows the course of posted prices from 1970 taking Arabian Light 34° and Libyan 40° as representative of the broader sweep of changes.

1.4 FUTURE PROSPECTS

The immediate consequences of these startling price rises are obvious enough. In the short run the demand for petroleum is very price-inelastic and this is particularly true for certain products such as petrol, for which no substitute is available.[16] Some American studies suggest that the price elasticity of demand for petrol might be as low as 0.13.[17] No detailed British studies of the price elasticity of demand for petroleum products have yet been published, but the recent reports by the National Economic Development Office and by the Central Policy Review Staff[18] both appear to anticipate that the demand for most petroleum products will be very price-inelastic in the short run. In the long run the position is more uncertain, but it is clear that a large reduction in the demand for petroleum products could be secured only by major changes in government policy and very high levels of investment in new energy or energy-saving technologies. The combination of a low price elasticity of demand for petroleum products and a high income elasticity of demand, so that over time the demand for petroleum products tends to grow more quickly than income as a whole, puts the oil-exporting countries in a very strong bargaining position. Higher oil prices will involve a huge transfer of income from oil-consuming to oil-producing countries. For example, even if we assume no increase of oil consumption on 1973 levels, the prices currently ruling would imply a current payments surplus for oil producers of about $75,000m matched by oil deficits of about $62,000m for the OECD countries and of $13,000m for the non-oil-producing developing countries. For the UK the oil deficit appears likely to be in the region of $8000m or some £3500m.

Higher oil prices have much the same impact on the economy as a rise in indirect taxes. In short, they will influence the pattern of consumption (the price effect) and tend to reduce the level of income (the income effect). The price effect seems likely to lead in the long run to a substantial reallocation of productive resources between alternative uses. For example, it has major implications for the transport industries and for other energy-intensive industries such as building materials, fertilisers and aluminium smelting and will produce a shift towards products that are less energy-intensive – smaller and lighter cars, diesel engines, etc. Even more important in the short run and medium run is the income effect of higher oil prices. To offset the deflationary effects on income and employment, the oil-consuming countries will have to adopt compensating fiscal and monetary measures to prevent a rise in unemployment, and in turn this implies the acceptance of large oil deficits on balance of payments current account. The financing of these deficits will necessitate close co-operation between nation states as the most obvious adjustment mechanism, increased imports by oil-producing countries as their incomes rise, will be severely limited by the small size and, therefore, the limited absorptive capacity of their economies.

In view of this a major recession can be avoided only if the oil-deficit countries are willing and able to borrow from the Arab states and thus finance their deficits on current account. Such borrowing, known euphemistically as 'recycling', faces many difficulties. The oil producers may be reluctant to keep their wealth in financial assets, or even in real assets ('bricks and mortar') whose realisation depends on the goodwill of other nations. The oil-importing countries may be reluctant to accept the political and strategic consequences of major investments in their assets by Arab states, and there is no automatic guarantee that such investment will be directed towards the nations in greatest need. The sheer volume of financial transfers required seems likely to create major uncertainties in world money markets. Recycling requires a broad measure of agreement between oil producers and oil consumers, and in turn between the oil consumers themselves, and neither historical precedent nor recent experience provides much assurance that the required degree of co-operation will be forthcoming.

Moreover, recycling only buys time – it makes it possible for the oil-consuming countries to avoid temporarily the need to deal with large deficits on current account. Borrowing may then be sensible in the short run, but it cannot proceed indefinitely on the scale made necessary by the recent rise in oil prices. If it did, present generations in the oil-consuming countries would be committing future generations to a heavy burden of international debt – a policy that is justified only if borrowing permits future generations a higher income than would otherwise be possible.[19] Borrowing for this, as for any other purpose, is only a *temporary* expedient; it cannot represent a long-run solution. In the long run only two solutions are possible – either the oil-producing countries must increase their imports from oil-consuming nations, or the oil-consuming nations must develop alternative sources of energy.

The short-run prospects for the oil consuming nations are bleak, but the longer-run position is much more complex and unpredictable. While consumers looking up at the new heights to which oil prices have risen may suffer from a severe pain in the back of the neck, producers looking down from the dizzy heights of current price levels, at some $10 per barrel for Persian Gulf oil landed in Europe, to the floor set by the minimum long-run production price at around 20 cents per barrel, may well suffer a bad attack of vertigo. It is difficult to know which condition will prove fatal.

The unprecedented rise in prices must produce two reactions: first, among those oil producers that need additional revenue, a desire to increase production because of the gulf between realised prices and the real resource costs of production; and second, among consuming nations, a search for cheaper, more secure, alternative sources of energy and investment in energy-saving technologies. A cartel is only as strong as its control over production, and as yet it does not appear that OPEC has reached any working agreement on the principles which will determine pro-rationing. It will be in the interest of Saudi Arabia and Iran to attach greatest weight to reserves; of Kuwait, the United Arab Emirates and Qatar to emphasise other 'economic' variables, particularly the oil income-to-government revenue ratio; and of Iran, Indonesia, Nigeria and Venezuela to emphasise population size and acreage. On the other hand, producers like Venezuela, with low reserves, and Libya, the United Arab Emirates and Qatar with small economies, and consequently low absorptive capacity, have a clearly defined interest in strictly restraining supplies in order to maximise present prices. This, together with OPEC's recent successes, may be the cement that holds the cartel together. Historically, cartels have proved to be unstable coalitions, but OPEC is the first cartel to consist of states that will have great difficulty in absorbing current revenues. The temptation to increase production will then be weak; while the recent record clearly demonstrates the advantages of solidarity.

The increase in oil prices will encourage a much more extensive search for new reserves, particularly outside the Middle East – in Alaska, the Athabasca tar sands and, increasingly, the continental shelf, pushing out into deeper waters as off-shore technology advances.[20] Coal, natural gas, hydroelectric and nuclear power have become more price-competitive and this will encourage increased supply. There are other more remote possibilities such as nuclear fusion, tidal power and solar energy. The difficulty is that, while the technical possibilities are considerable and varied,[21] it is much more difficult to predict whether and when any alternative source of energy will prove an economic substitute for oil. It is probable that a continued rise in energy consumption over the rest of the 1970s can only be met, as in the past, by increased oil production, and a quite new factor will emerge as the US increasingly becomes a major net importer of oil.

From this it would appear that the oil producers may be able to sustain prices at extremely high levels over a period of some years. Certainly, a major shift from oil to other sources of energy would require extremely heavy capital expenditure, which will be undertaken only if the rise in oil prices is regarded as permanent. For fuel oil, and indeed for all other petroleum products, the price elasticity of demand will be greater the longer the period of time that elapses and the greater the certainty that any increase in prices will prove permanent. Given time, the price mechanism will produce a powerful stimulus towards energy conservation and away from energy-intensive activities. This will tend to reduce oil prices, or at least to restrain further substantial price increases.

1.5 CONCLUSIONS

The temptation which oil producers might feel to break ranks and, more importantly, the evident incentive for oil importers to develop new sources of energy may in time produce a fall in the real price of oil. However, this is by no means certain and the floor to oil prices, set by the long-run supply price of alternative fuels, does not seem likely to be less than $7 a barrel (at 1973 prices). Any return to an era of 'cheap' oil is highly improbable, but beyond this considerable uncertainty must remain.

Nobody knows whether the supplies of other forms of energy (coal and indigenous oil) will pursue their most likely path. Nobody knows whether consuming countries will be wise enough or clever enough to concentrate their cut in purchases on OPEC oil rather than on indigenous fuels. And nobody knows how cleverly and strongly OPEC's strategy will be conducted. So in my view a certain ambivalence and caution must be exercised by the decision makers under present circumstances.[22]

However, whatever the long-run equilibrium price of oil, it is clear that the power of the oil-exporting nations to enforce an extremely high price for a substantial period is much greater than previously anticipated. The best defence is to develop alternative sources of energy. This underlines the importance of the North Sea oil and gas reserves, for they are of immense economic value at current price levels, or indeed at prices even substantially below present levels.

Yet, 'waiting for North Sea oil' does not constitute an adequate policy stance. By the late 1970s North Sea oil will eliminate the oil deficit and hence the need for continued massive overseas borrowing. It can help produce a strong balance of payments position on current account and this, in turn, will permit more expansionary economic policies than would otherwise have proved possible. Nonetheless, North Sea oil cannot entirely insulate the UK from the effects of a major world depression. For this reason the UK must continue to pursue policies that recognise the need for concerted action to deal with the oil deficits arising from the rise in crude oil prices.

Finally, it is prudent to take all available measures that will reduce the need for oil imports in the next few years. At present there is no indication that this is understood. On the contrary, the prospect of North Sea oil seems to have convinved UK governments to ape the Eskimos who 'became so convinced of the imminence of the millennium that they stopped hunting and ate into their stores of food'.[23] So, North Sea oil is 'sold forward' or 'mortgaged' (the terminology is a matter of choice) against present borrowing to finance the oil deficit. Such a passive policy is not enough. Fiscal weapons must be ruthlessly directed towards strengthening the market forces which will themselves tend to reduce oil consumption, and hence the need for overseas borrowing. A differential road licence tax discriminating against cars with poor petrol consumption, and building regulations requiring improved insulation for new dwellings, are two examples of a battery of available weapons. Above all, the price of energy should not be kept artificially low. Such measures should be implemented in any event. They become even more urgent viewed against the possibility, however remote it might appear at present, that the UK may have to borrow heavily against high oil prices, only to find prices falling by the late 1970s when indigenous oil becomes available in quantity. The North Sea oil and gas reserves are a gift to be nurtured, not squandered in an attempt to avoid harsh realities.

NOTES TO CHAPTER 1

1. C. Tugendhat 'Energy Policy' *Economic Studies* (October 1969) p. 80.
2. Statistics taken from *Oil: The Present Situation and Future Prospects* Organisation for Economic Co-operation and Development (1973)
3. Producers are required to 'land' North Sea oil in the UK, but there is no legal restraint on their right subsequently to export crude or refined products. Indeed, any such restraint would be contrary to the competitive rules adopted by the EEC.
4. The best accounts of British fuel policy are M. V. Posner *Fuel Policy: A Study in Applied Economics* London, Macmillan (1973) and G. L. Reid, K. Allen and D. J. Harris *The Nationalised Fuel Industries* London, Heinemann (1973).
5. Additional factors were the growth of the 'independents' leading to increased competition, and economies of scale in refining and in tanker shipping.
6. See *Oil: The Present Situation and Future Prospects,* op. cit. pp. 276–7.
7. The convention in the oil industry is to express prices in dollars per barrel. We have followed this convention, but in discussing the effects on the British economy we have converted into sterling at an exchange rate of $2.30 : £1.
8. These data are taken from M. A. Adelman *The World Petroleum Market* Baltimore, Johns Hopkins UP (1973).
9. ibid.
10. ibid. p. 77.
11. Although there has been severe price competition in 'uncontrolled' markets – international aviation and bunkers, public tenders, major new projects, independent refinery tenders etc.
12. See M. A. Adelman 'The World Oil Outlook' in M. Clawson (ed.) *Natural Resources and International Development* Baltimore, John Hopkins UP (1964).
13. See Z. Mikdashi *The Community of Oil Exporting Countries* London, Allen & Unwin (1972); F. Rouhani *A History of O.P.E.C.* London, Praeger (1972); and G. W. Stocking *Middle East Oil* London, Allen Lane (1971).
14. Mikdashi op. cit. p. 139
15. *Petroleum Economist* (February 1973) p. 42.
16. Oil supplies 99 per cent of the energy requirements of the UK's transport industries.

17. L. W. Weiss *Case Studies in American Industry* (2nd edn) Chichester, Wiley (1971) p. 255.

18. *Energy Conservation* London, HMSO for Central Policy Review Staff (1974) and *The Increased Cost of Energy: Implications for UK Industry* London, HMSO for the National Economic Development Office (1974).

19. Moreover, the margin of additional future income must be sufficient to service the debt on current borrowing. It follows from this that borrowing that results only in maintaining current *consumption* is indefensible, whereas borrowing that permits, directly or indirectly, a higher level of investment may be justifiable.

20. Higher prices will also encourage more intensive production of proven reserves. At the prices ruling in the past it has proved economic to recover only some 40–50 per cent of existing deposits, but more expensive production techniques will now prove economic and the recovery rate may rise significantly.

21. See *Energy Conservation,* op. cit.

22. M. V. Posner 'The Energy Price Rise and World Trade and Payments' (mimeograph) Manchester, Association of University Teachers of Economics (1974).

23. P. Worsley *The Trumpet Shall Sound* London, Paladin (1970); quoted in K. J. Alexander 'The Political Economy of Change' *Presidential Address, Section F., Economics, British Association for the Advancement of Science* (1974).

CHAPTER 2

Government Policy at Sea

Enter ye in at the strait gate: for wide is the gate, and broad is the way, that leadeth to destruction, and many there be which go in thereat:
Because strait is the gate, and narrow is the way, which leadeth unto life, and few there be that find it. (Matt. 7, vv. 13–14)[1]

2.1 INTRODUCTION

The pace and the extent of the North Sea search for oil and gas, which has been proceeding in the UK sector since the mid-1960s, is strongly influenced by government policy. Under international law, the UK government has the exclusive right to exploit the natural resources of its continental shelf. The conditions which it establishes regulate the rate of exploration and production at sea, and these in turn determine the impacts on income and employment ultimately felt on land. The system of exploration and production licensing adopted, and the financial regime applied to the production of oil and gas, are then the centrepoint of British oil policy. On this all else depends. If these aspects of policy are at sea, metaphorically as well as literally, then the situation cannot be retrieved by any other combination of policies.

It will be argued in this chapter that the policy framework established with regard to North Sea activities has been more successful than is commonly realised. It has thus far ensured a high rate of exploration and exploitation, and some of the 'loopholes' to which attention has been drawn have been more apparent than real. However, many of the crucial economic decisions, relating to pricing, taxation and production policy, have yet to be made and on these the ultimate success of British policy will turn. This delay is in some respects understandable as the nature of the problem has been transformed

18

by the rise in the price of crude oil. It is difficult to believe that any financial regime which might have been established in the early 1970s would be appropriate in current circumstances. Yet undue delay will, by creating additional uncertainty in an area where risks and uncertainty are always high, have an adverse effect on exploration and production, and the broad lines of British policy must soon be established.

This chapter discusses the background against which pricing, taxation and production policy decisions will be taken. It looks at the development of British policy and at the policy options that are now available. We shall discuss the legal framework within which Britain exploits mineral deposits in the North Sea, the controls over exploration and production activities, the taxation regime that is likely to be appropriate and the principles that should establish the rate of oil and gas production.

2.2 THE LEGAL FRAMEWORK

The North Sea search was fathered by the discovery, at Slochteren in Holland, of the second largest gas field in the world. But exploration drilling had to await an appropriate legal framework. From 1934 the Crown has enjoyed exclusive property rights over all natural gas and oil found on shore and within UK territorial waters. Similar rights were exercised by other governments, but the sea and the seabed outside territorial waters were ungoverned, although subject to opportunistic raids by national governments seeking to extend their right to explore and exploit the resources of the sea and the seabed. Claims over territorial waters varied from three miles in Australia to two hundred miles in the case of some Latin American countries. Fishing limits claims often followed the same boundaries, but were sometimes for wider areas (e.g. Iceland). Finally, with improved off-shore technology, and hence the growing prospect of industrial and commercial advantage, a number of countries (most noticeably the UK and Venezuela over the Gulf of Paria in 1942 and the USA over its continental shelf in 1945) had laid claim to the natural resources of the seabed adjacent to their coasts. Indeed, the Truman Proclamation of 1945 was the catalyst to future developments leading to the 1958 Geneva Continental Shelf Convention.

Unilateral action could hardly have constituted a basis for petroleum exploration in the North Sea given that it is bounded by seven littoral states: the UK, Norway, Denmark, West Germany, the Netherlands, Belgium and France. However, the Geneva Convention, while leaving some major issues unresolved and confusing others yet further, did provide a basis for a division of competence in the North Sea. The Convention extended the sovereign rights of the littoral states to include the exploration and exploitation of the natural resources of the seabed on the continental shelf to a depth of two hundred metres or 'to where the depth of the superjacent waters admits of the exploitation of natural resources'.[2] The extent of each state's rights depended upon its reaching agreement with other bordering countries, and given mutual agreement the boundary line could be determined by any set of principles. However, failing such agreement Article 6 of the Convention established the principle of equidistance to guide countries that were partners to the Treaty. Thus:

> In the absence of agreement, and unless another boundary line is justified by special circumstances, the boundary shall be determined by application of the principle of equidistance from the nearest points of the baselines from which the breadth of the territorial sea of each State is measured.

Between two coastal states, such as the UK and Norway, the boundary line is the median line between them and indeed the line demarcating the UK sector of the North Sea is the result of five separate agreements each of which is consistent with the principle of equidistance. The effect of this criterion can be seen from the frontispiece, which shows the national sectors determined for the North Sea.

There remain knotty questions of interpretation and some of open dispute between countries – for example, the fierce disagreement between Greece and Turkey over their respective rights in the Aegean Sea and the recurring battles between South Vietnam, China and North Vietnam over islands in the South China Sea. In the Atlantic and the Irish Seas there appears to be ample scope for dispute between the UK and Ireland, despite the UK's attempted assumption of sovereignty over Rockall and its belated discovery of a stretch of water known as the Celtic Sea, a term previously unknown to geographers!

In the North Sea itself a dispute between Denmark, the Netherlands and West Germany was settled in favour of the latter when the Court of Justice in the Hague determined that the boundaries should be settled by agreement based on 'equitable principles'. As far as the UK was concerned the major issue was whether the Norwegian trench, with a depth well in excess of 200 metres, constituted the limit to Norway's continental shelf, a question settled in Norway's favour by an agreement in 1965 which ignored the trench in determining the median line between the two countries. The Geneva Convention had been ratified by a minority of only forty countries as late as 1970, and the clause limiting the powers of coastal states 'to where the depth of the superjacent waters admits of the exploitation of natural resources' has been overtaken by the march of off-shore technology. So much so that it no longer sets any ultimate limit to the possible rights of coastal states over all the seabed and is therefore likely to be contested by land-locked and shelf-locked states. However, the Convention did provide a framework for exploration and exploitation in the North Sea and the UK Continental Shelf Act of 1964 claimed for the UK 'all rights exercisable' over oil and natural gas deposits on its continental shelf. The estimated area of each country's sector, its share of the total North Sea area and its population is shown in table 2.1.

TABLE 2.1 ESTIMATED AREA OF NORTH SEA JURISDICTION AND POPULATION OF BORDERING COUNTRIES

	Area – sq. miles ('000)	% of total	Population ('000)
UK	95.3	46.7	55.9
Norway	51.2	25.1	3.9
Netherlands	21.8	10.7	13.3
Denmark	18.8	9.2	5.0
West Germany	13.9	6.8	61.7
Belgium	1.6	0.8	9.7
France	1.6	0.8	51.7
	204.2		

Sources: Royal Scottish Geographical Society *Scotland and Oil* (1973); *The OECD Observer* (February 1974)

The principle of equidistance favours countries with long coastlines and particularly those, like Norway, with a long coastline and a small population. The UK sector covers almost half of the North Sea area

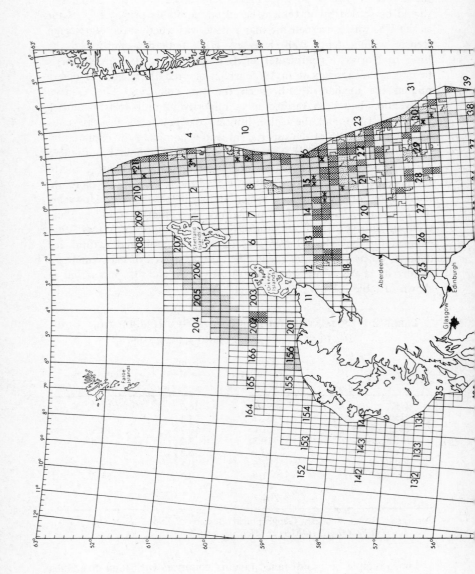

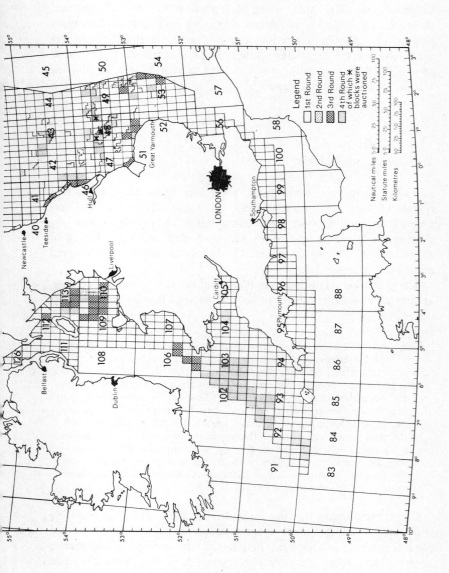

and the Continental Shelf (Jurisdictional) Order of 1968 further subdivides the UK area, along the parallel 55° 50′ North, into Scottish and English areas where Scots law and English law apply respectively. Where the terms 'Scottish' and 'English' waters are used subsequently they refer to these areas so defined. The Scottish area is 62,500 square miles and the English area 32,800, so that the UK sector divides in the ratio 2:1.

2.3 LICENSING POLICY

Following from the Continental Shelf Act there have been four rounds of licensing (in 1964, 1965, 1970 and 1971–2) which have granted exploration and production rights to licensees. For the purposes of licensing, the North Sea is divided into equal blocks of approximately 100 square miles. The blocks allocated in each round are shown in figure 2.1.

The policy considerations which dictated the first round of licensing, and which have been subject to only minor qualification subsequently, were:

> First, the need to encourage the most rapid and thorough exploration and economical exploration of petroleum resources on the Continental Shelf. Second, the requirement that the applicant for a licence shall be incorporated in the United Kingdom and the profits of the operation shall be taxable here. Thirdly, in cases where the applicant is a foreign-owned concern, how far British oil companies receive equitable treatment in that country. Fourthly, we shall look at the programme of work of the applicant and also at the ability and resources to implement it. Fifthly, we shall look at the contribution the applicant has already made and is making towards the development of resources of our Continental Shelf and the development of our fuel economy generally.[3]

Quick exploitation was the main driving force of British licensing policy and from this many other consequences flow. The perceived need for rapid development determined the method of allocating licences,[4] the reliance on foreign capital and the financial terms initially established. The successful applicants were not determined by a competitive auction where the licence went to the highest bidder, but by administrative discretion based on the proposed work programmes of the applicant. Other things being equal, the applicant with the most intensive work programme was successful. Rapid exploration was further encouraged by requiring each licensee to surrender one-half of his area after a six-year period, so that there was an obvious incentive to determine which part of any area offered the best commercial prospects.[5] Finally, the annual rental of each block licensed was low. In the first round it was fixed at £6250 per annum for the first six years, rising subsequently in £10,000 stages to reach a maximum of £72,500 per year, and although the rent was raised subsequently, in the third and fourth licensing rounds, it remained almost nominal. In this way, it is argued, only a minimum strain is put on the working capital of the oil companies and this enables them to finance rapid exploration.

Speed dictated a reliance on foreign-controlled oil companies, particularly American companies. It was considered that the two UK, or partly UK, majors, British Petroleum and Shell, did not have the necessary resources to develop all the licensed areas as quickly as was desired, but some preference was shown to UK companies, as they appear to have obtained a relatively higher proportion of the more promising areas (see p. 76).

In the second licensing round (1965) the new Labour administration added two further criteria. First, it would consider the contribution which applicants had made to the UK balance of payments and to creating employment in the United Kingdom with particular reference to regional considerations and, secondly, 'proposals which may be made for facilitating participation by public enterprise in the development and exploitation of the resources of the Continental Shelf'[6] would be taken into account. The effect of this latter criterion is clearly shown in table 2.2.

Over the four rounds, the share of UK interests tended to rise. The exceptions to this are those blocks auctioned by competitive bidding in the fourth round, which again suggests that some administrative discretion was exercised in favour of UK interests in the earlier licensing rounds. Nonetheless, the share of foreign companies remained high in all four rounds. Foreign interests never accounted for less than 62.5 per cent of the areas licensed and in this American interests predominated. We have calculated that, at November 1974, American-based companies controlled 38 per cent of the estimated production potential from established North Sea oil fields, compared

TABLE 2.2 PERCENTAGE OF LICENSED TERRITORY HELD BY BRITISH INTERESTS

Round	Gas Council	Public sector National Coal Board	HM Govt*	Total public sector	Private British interests**	Total British interests
1	4.5	2.1	2.6	9.2	13.5	22.7
2	7.6	4.5	3.4	15.5	18.1	33.6
3	10.1	4.5	5.4	20.0	16.5	36.5
4 (auction)	2.1	6.2	1.7	10.0	10.0	20.0
(discretionary)	3.0	2.3	4.3	9.6	25.1	34.7
All rounds	5.0	3.0	4.0	12.0	20.0	32.0

*Calculated by taking the Government's share of BP shares, 48.6 per cent.

**Taking Shell as 40 per cent 'British'.

Source: First Report from the Committee of Public Accounts, *North Sea Oil and Gas* (1972–3).

with 44 per cent controlled by UK-based companies and 18 per cent controlled by companies of other nationalities.[7]

Table 2.3 shows details of the number of blocks offered and applied for, the number of production licences granted and the total area licensed in each of the four licensing rounds.

TABLE 2.3 AREAS LICENSED AND PRODUCTION LICENCES GRANTED

Round	Blocks offered (no.)	Blocks applied for (no.)	Production licences granted (no.)	Total area licensed ('000 square miles)
1	960	394	53	32
2	1102	127	37	10
3	157	117	37	8
4	435	286	118	24

Source: First Report from the Committee of Public Accounts, *North Sea Oil and Gas* (1972–3)

There were two major licensing rounds. Round 1 was heavily concentrated in the southern North Sea, the English sector as defined on p. 24, reflecting the geological appraisal following the Slochteren discovery. Rounds 2, 3 and 4 were more heavily concentrated in the northern North Sea, the Scottish sector as defined on p. 24, but also tidied up some of the areas unallocated in the southern sector and leased blocks west of the Shetlands and in the Celtic Sea.[8] Of these rounds, the fourth round was the most important, resulting in a total area licensed of 24,000 square miles.

In the first three rounds all licences were allocated by administrative discretion. Competitive licensing was not attempted until the fourth round and even then was restricted to only 15 of the 286 blocks eventually allocated. The arguments for discretionary licensing can be quickly summarised. The UK was heavily dependent on external sources of fuel and it was desirable to provide a secure and, hopefully, cheap source of fuel from indigenous sources, which would benefit the balance of payments. The North Sea offered such a possibility, but it was an unproved area which could be tested only by expensive exploration and drilling carried through at the limits of existing off-shore technology. The licensing system adopted allowed the Department of Trade and Industry to exercise its discretion in favour of companies that would pursue an active search programme.

It had the further alleged advantage that it allowed some (concealed) discrimination in favour of UK operators and overt pressure could be placed on all operators, UK or foreign, to buy British goods and services for off-shore exploration and production.

There is little doubt that the major objective was met. Exploration has been more rapid and more extensive in the UK sector of the North Sea than in any other national sector, although a major contributory factor has been the high success rate achieved in UK waters. By as early as 1970, only four years after exploration activity had begun in earnest, total proven reserves of natural gas were 27 million million cubic feet, equivalent to a daily production flow of between 3500 and 4000 million cubic feet per day (m.c.f.d.) over a twenty-five-year period.[9] The estimated 1975 supply of natural gas was approximately three times total gas consumption in 1968, by which year it was estimated that natural gas would meet 14 per cent of UK fuel requirements.

Given Britain's dependence on external fuel supplies, a dependence underlined by the rise in the price of Middle East oil, it is clear that speedy exploitation was extremely important. This the discretionary system did achieve. Speed was almost certainly at the price of some short-term loss to the Exchequer, as a competitive system of licensing would almost certainly have resulted in a higher annual rental per block licensed. However, given the very imperfect knowledge of the geological conditions prevailing and the limited experience of operating in North Sea conditions, the potential loss of immediate revenue could not have been absolutely large, and was relatively extremely small compared with the gains of quick exploitation.

While this appears to be the correct assessment for the first three licensing rounds, it is difficult to resist the conclusion that the continued heavy reliance on the discretionary system in the fourth round was mistaken. It could be defended on the grounds that the increase in the posted prices, levied by the OPEC countries in September 1970, further underlined the need for haste; and there was also concern at what appeared to be a fall-off in exploration activity from 1970 in consequence, it would appear, of a rising dry-hole ratio.[10] On the other hand, the major Ekofisk oil strike had been announced in November 1969. In the UK sector the discovery of the Forties field was announced in 1970, the more minor discoveries of Argyll and Auk in 1971, and in the Brent field the discovery well of what later turned out to be a major field had been drilled before the fourth round. Certainly, when the fourth licensing round was completed the results indicated that the oil companies and the Department of Trade and Industry (DTI) had very different assessments of likely prospects. The 271 blocks allocated through the discretionary system brought an income of only some £3m. The handful of fifteen blocks, which had been deliberately selected by the DTI as a 'representative cross-section' of the whole area to be licensed, was sold by competitive auction for £37m. In short, the DTI took a conservative, even a pessimistic view of the future, while the oil companies were evidently more optimistic. Subsequent events have proved the assessment of the oil companies to have been correct.

The Public Accounts Committee of the House of Commons considered that the decision to rely so heavily on discretionary licensing in the fourth round was a major commercial misjudgement on the part of the DTI. That it was a misjudgement we do not doubt, but its importance was greatly exaggerated by the Public Accounts Committee. In the first place, it is quite mistaken to believe that the price obtained for the fifteen blocks auctioned could have been sustained for a much larger number of blocks. No less than £21m of the total sum of £37m realised through the auction was obtained for the so-called 'Golden Block' (block 211/21), whose subsequent history belied its name. Most important of all, the sums involved were of minor significance compared with the gains of quick exploitation, which was secured by DTI policy, or with the revenues that will ultimately accrue from an appropriate financial regime for North Sea operations.

Viewed in this light the issue hardly merited the righteous indignation displayed by the Public Accounts Committee.[11] Yet the case for a heavy reliance on discretionary licensing is much weaker than the DTI's evidence would suggest. The most powerful argument in its favour is that a competitive auction would be unfavourable to British interests, and this does seem to be borne out by the experience of the fourth round where British interests obtained a relatively low share of the blocks sold through auction. However, concealed discrimination, because it does not have to be justified, tends to be inefficient and there are strong *a priori* grounds for preferring a system of allocation which places more weight on the market solution and less on administrative discretion.

Some of the further advantages claimed for the discretionary system are even more difficult to sustain. In particular, there is no reason to believe, as the oil companies claim, that a system of auctioning licences would prohibit rapid exploration because it would significantly reduce their working capital. After all, if the oil companies were concerned about this issue it would surely be reflected in the lower prices they would be prepared to bid! The State's interests could be protected by specifying the minimum work programme required on each block. Indeed, auctioning might lead to faster exploitation. After all, having invested capital in buying a licence the rational licensee will wish to obtain a return on that capital as quickly as possible. There is no reason to suppose that this elementary point has escaped the oil companies.

2.4   THE FINANCIAL REGIME

The financial regimes applied to natural gas and to oil seem likely to be substantially different. It is generally held that the regime applied to natural gas is simple and effective in ensuring that private companies do not make excessive profits on established fields. This is so, but it is also extremely inefficient. On the other hand, while there has been considerable public criticism about the taxation system likely to apply to oil fields, it must be borne in mind that the details of any system have not yet been established and, as yet, private companies have incurred substantial costs with no return. There is every possibility that the fundamental weakness in British policy will arise from the peculiar regime applied to natural gas, although it is true that, given the importance of the oil fields, the potential difficulties are greater for oil.

The financial regime applied to natural gas consists of vesting the British Gas Corporation with monopoly powers to purchase natural gas found in the UK North Sea. The licensee is required by law to offer gas for sale to the Gas Corporation at a 'reasonable price', and the Minister of Power (now Minister for Energy) is the final arbiter as to what constitutes a 'reasonable price'. Further, the Gas Corporation has also acquired, consequent on the Gas Act of 1965, powers to supply gas to the area boards, so that it effectively has monopoly powers over purchase price, distribution and final price.

This monopoly power enables the Corporation to keep down the purchase price of natural gas and successfully prevents private companies from enjoying excessive profits. The difficulty is to ensure this, while maintaining sufficient incentive to encourage further exploration. The stated aim of government policy is to establish a 'reasonable price', which is as low as possible while being consistent with the principle of encouraging further exploration. This balance is not an easy one to strike and it is difficult to dispute the contention that a satisfactory balance has not been established. The purchase price, although it initially guaranteed an adequate return on large fields, has been kept at a level that discourages further exploration and the exploitation of some smaller fields in the southern North Sea, which are economically viable given the prices ruling for other forms of energy.[12] Moreover, political pressure has kept the final price to the consumer artificially low. This results in a wasteful use of natural gas, instead of confining it to premium markets where it has distinct advantages over other fuels. The effects of this type of policy are already clearly visible in the USA, where the artificially low price established for natural gas has allowed it to penetrate markets which would be better served by other fuels, while discouraging exploration for new sources of supply. If UK policy continues along present lines, and there is every indication that it will, then it will inevitably produce the same inefficient use of energy and a lower supply of energy than a realistic and flexible purchasing policy would make possible.

It is fortunate that this particular 'solution' is unlikely to be applied to oil. Instead the initial principles which were established for oil production were as follows.

1. As we have seen, there was a low annual rental for each block which was fixed in price for the first few years and increased thereafter to a stipulated maximum.

2. The Exchequer was to receive $12\frac{1}{2}$ per cent of the wellhead value of any petroleum produced.

3. Profits were subject to UK tax.

It was officially estimated that the government's share of profits from North Sea operations would be in the range of 50–60 per cent, comparable with estimated returns under the financial regimes established by the Netherlands and the USA, but less than Norway (55–80 per cent), Nigeria (70–75 per cent) and the Middle East states (75–80 per cent).[13] The UK had aimed for a return lower than that obtained by most oil producers, again on the grounds that rapid exploration should be encouraged, but the Public Accounts Committee of the House of Commons believed that even this low target rate of return was falsified by a crucial weakness in the taxation system adopted.

The Committee argued that, while in principle the system appeared water-tight, in practice it had the porosity of a sieve. We believe this view was mistaken, but as it has been widely accepted as displaying the incompetence of UK policy it merits more detailed examination.

The estimated government share rested on the assumption that profits earned on North Sea operations would be subject to UK corporation tax. This was guaranteed, or so it was thought, by the second condition laid down in the first licensing round, namely that 'the applicant for a licence shall be incorporated in the United Kingdom and the profits of the operations shall be taxable here'. This appeared to shut the stable door, but the Committee considered that the horse was already outside. The oil majors had accumulated 'losses' on their Middle East operations which, it appeared, they could offset against profits made in the North Sea. If so, the UK oil majors would probably not have incurred any liability to UK tax before 1980 at the earliest.

This possibility arose as follows. As we have seen (chapter 1), the Middle East countries levy taxes on the posted price of oil. The posted price is a tax reference price and is in no sense the market price for oil. Yet the use of the posted price to transfer oil from the producing to the trading company of an oil major could produce tax 'losses' which the oil companies could then offset against profits earned elsewhere. Thus, the oil majors' total UK corporation tax liability amounted to only £½m between 1965 and 1973.

A simple example may help to illustrate how the system works. Suppose Middle East countries levy tax at a marginal rate of 50 per cent and that they then increase posted prices by $2 a barrel. The tax paid to the producer country will rise by $1 a barrel. Suppose further that the oil major transfers oil from its production company to its trading company at the new posted price ($2 higher), but also changes the final price by $1 a barrel to offset the higher tax paid to the producer country. Then the end result will be that the company's overall profit rate will not fall, but it will carry forward a further $1 per barrel in tax 'losses' from its trading operations to offset against profits earned elsewhere. As a result of this process it was estimated that the nine oil majors had accumulated 'losses', which could be offset against North Sea operations, of no less than £1500m.

Had this been overlooked it would indeed have had serious consequences. However, even before the Public Accounts Committee 'uncovered' this difficulty the Inland Revenue had taken up the issue with the UK oil majors. There was never any serious prospect that the majors would have been allowed to use these 'losses' to offset against North Sea profits. Moreover, it was also possible, given the terms of the initial licences, to levy a special rate of tax on North Sea operations. Thus the initial terms only stipulated that UK taxes would apply, but they did *not* specify that the rate of tax should be the same as that applied to other commercial operations. This is a very important point given the increase in overall oil prices from 1973. These price increases would have provided the oil companies with huge windfall gains if they were only subject to UK corporation tax. However, as the initial terms did not specify the rate at which North Sea oil profits would be taxed, they provided much greater flexibility than has often been supposed. The Department of Trade and Industry (now the Department of Energy) has often been criticised,[14] possibly with some justification, for its slowness in announcing the need for a new system of taxation for North Sea oil operations, but it cannot fairly be criticised for providing a system which could not be adjusted in the light of changed circumstances.

The above argument implicitly assumes that a unique financial regime should be established for North Sea oil and gas operations. The justification of this is straightforward. It is that the price that is just sufficient to induce exploration and production of oil and gas is much less than the market price of the commodity. The difference between these prices measures the 'economic rent' that accrues to the owner of the natural resources. Originally the owner is the State, but the process of licensing confers temporary property rights for the exploration and production of oil to the licensee. Therefore, the function of any financial regime established for petroleum exploration and production is to ensure that the economic rent accrues to the Exchequer rather than to the licensee. The producer should obtain terms which guarantee him a payment equivalent to his transfer earnings, i.e. a payment just sufficient to induce the producer to continue to employ his productive resources in their current use rather than transferring them elsewhere.

Ricardo defined pure rent as the payment due to the 'original and indestructible powers of the soil'.[15] The supply of land was fixed and it was assumed that it had only one use (growing corn). The landlord would prefer to obtain a rent from this land rather than leave it idle. The return he received was the rent, i.e. a surplus above that necessary to retain the land in its present use. The amount of this rent depended on the intersection of the demand schedule with a completely inelastic supply schedule. In figure 2.2(a), all returns to the land, measured by the shaded area, are pure rent. Tax-paying capacity is, in principle, equivalent to the shaded area. In other words, the shaded area can be captured through taxation without affecting the supply of land, although considerations of equity may require some compensation to the landlord who has bought the land on the expectation of receiving rent income in the future.

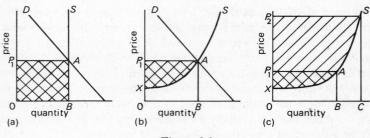

Figure 2.2

Natural resources, unlike land, are not indestructible or fixed in supply. They are finite, and hence exhaustible, and their supply can be increased through greater expenditure on exploration and on improved methods of recovery. However, modern economic theory suggests that the payments to most factors of production (land, labour or capital) comprise two elements – an economic rent and transfer earnings. For example, in figure 2.2(b) the shaded area XP_1A represents rent and the unshaded area $OXAB$ represents transfer earnings.

To illustrate the argument in the case of North Sea oil deposits, let us assume in figure 2.2(c) that the supply schedule XAS represents the minimum prices necessary to induce the exploration and production of different quantities of oil. The schedule slopes sharply upwards, representing the higher costs of finding and exploiting marginal fields in North Sea conditions. We can envisage low costs of production from large oil fields, found with little exploration, in shallow water and close to a convenient land fall, and high costs of production where the reverse conditions apply. As North Sea oil reserves represent only 2 per cent of known world reserves,[16] we can assume that they have little influence on world prices. We can take oil prices as determined outside our system and here we simply assume, for the purpose of illustration, that OP_1 represented the price of crude oil in 1971–2 at the time of the fourth licensing round. In these circumstances the aim of British financial policy can be represented as attempting to establish terms which would guarantee the producer $OXAB$, equivalent to transfer earnings, while ensuring that the economic rent XP_1A accrued, in one form or another, to the Exchequer. As figure 2.2(c) clearly shows, this would require a system of sufficient flexibility to allow for very different costs of production from the North Sea. In other words, the economic rent would be high on low-cost production and low on high-cost production. Further, the system should also be capable of dealing with a situation of a major change in oil prices. This is represented in figure 2.2(c) by the increase in price from P_1 to P_2, the relative size of the rise being similar to that which actually resulted in 1973 from OPEC activities. In the illustration the increase in price from P_1 to P_2 increases economic rent to XP_2SA and transfer earnings to $OXSC$. **1892042**

The *minimum* aim of British policy should be to remove the windfall gains associated with recent rises in world oil prices, i.e. the additional area of economic rent illustrated as P_1P_2SA. However, this still leaves an open question as to what type of financial measures should be adopted. Even in broad terms we can conceive of seven different types of taxation regime. The question we must consider is this: which type of regime is likely to be most efficient in capturing the economic rent for the original owner, the State, while still guaranteeing a satisfactory rate of exploration and production?

The seven possible systems of control are:

1. outright nationalisation of the oil industry;

2. the establishment of a monopoly buyer for oil to parallel the activities which the Gas Corporation carries through for natural gas discoveries;

3. the sale of licences through competitive auction;

4. a sales tax, such as royalties which are presently levied at $12\frac{1}{2}$ per cent of the wellhead value of crude oil;

5. an output tax, for example a barrelage tax, which would vary with the volume of output;

6. corporation tax together with an 'Excess Profits Tax' levied at, say, 75 per cent;

7. state participation through a holding in each oil field.

All these methods of control could be combined with each other in various ways, save the first two. It would be difficult, in all equity, to nationalise existing licence-holders without very heavy compensation. Moreover, outright state ownership and control, while in theory offering the simplest solution for ensuring that the State captures any economic rent, is quite impractical.[17] It overlooks the extent to which off-shore technology and engineering know-how is dominated by the US operating companies. Nationalisation would deal a crippling blow to the rate of exploration and exploitation. The technological dominance of US companies is a present fact and UK policy, however constructed, must rest on the principle that it has to be harnessed to serve Britain's economic needs. Or, to put it another way, private business interests, including American business interests, must be encouraged to participate in exploitation and production and this, in turn, requires that they should be able to obtain a rate of return on their capital at least equal to the return that they would be likely to earn in the best alternative use.[18]

The establishment of a monopoly buyer for oil would also have severe disadvantages. Such a policy can be followed with natural gas because the monopoly buyer, the Gas Corporation, also has control over distribution and market outlets. With oil, the monopoly power of a single buyer would not result in lower prices to the consumer unless it was accompanied by the nationalisation of all downstream activities. Moreover, unlike natural gas, which is used in the domestic market, most North Sea crude is likely to be exported. Finally, a monopoly buyer with an inflexible purchasing policy would inevitably discourage production from more marginal fields, just as the purchasing policy pursued by the Gas Corporation in the southern North Sea has discouraged further exploration in that area where some smaller gas fields could be profitably exploited given current energy prices.

We have suggested above that a larger element of competitive auctioning should have been introduced in the fourth round, and we would certainly expect less reliance on discretionary licensing in subsequent rounds. However, although there are grounds for supporting *some* significant movement in this direction, we do not believe that competitive auctioning provides a sufficient guarantee that the State will capture the economic rent. Using standard economic theory, it is easy enough to demonstrate that an auction would remove all the economic rent where there was perfect foresight, a very large number of bidders and blocks, and the absence of any collusion between the bidders.[19] It is easier still to demonstrate that none of these conditions hold in the North Sea. Given imperfect knowledge, a relatively small number of bidders and the possibility of collusion, it is likely that the price obtained through competitive auctioning would not be sufficient to extract the economic rent from the producers who obtained the larger and more profitable finds. It is significant that no country has ever relied exclusively on such a system, although in theory it has much to commend it. There is still a case for competitive auctioning on the simple ground, supported by *a priori* argument and practical experience, that a number of competing private interests are more likely to arrive at a competitive market price for a block than will DTI relying on administrative discretion. But the argument cannot be extended much beyond this and certainly British policy cannot rely solely, or even mainly on competitive licensing.

The other possibilities – a barrelage tax, royalties, excess profits tax and state participation – are much more serious candidates and, indeed, the present Labour administration appears committed to some combination of the latter two measures. Discrimination between the alternatives requires some view as to the nature of the fiscal problem: specifically, what rate of profit can be expected and how might this vary from field to field? It is only when we have formulated such a view that we can devise an appropriate strategy. Accordingly, we shall look at the capital costs, operating costs and revenues of the four oil fields for which reasonable estimates can currently be made – Auk, Argyll, Piper and Forties.

These fields offer an interesting cross-section for they include two small fields, Auk and Argyll, with low output *and* low capital costs, the medium-sized Piper field and the giant Forties field, both of which are heavily 'front-loaded', requiring very substantial capital expenditure well in advance of any revenue stream. This variation in operating conditions will prove typical of the North Sea and it poses problems for any fiscal system. The trick is to ensure that the State captures the economic rent without discouraging further exploration and production. This is extremely *difficult* to achieve and any simple system of taxation which applies uniform conditions will certainly fail to meet one of these two objectives.

In Appendix 2.1 we have set out the estimated capital and operating costs for the four fields and also the probable rate of production. At this stage we can only estimate likely costs and output, but we have chosen the most detailed and careful estimates available and have adjusted them upwards[20] to take into account the latest developments (to November 1974). All these estimates simply reflect the best view which can be taken, given that current knowledge and the actual outcome must depart, in lesser or greater degree, from the figures put forward. The greatest difficulties lie with the smaller fields as it is particularly difficult to establish accurate estimates of their potential in advance of the commencement of production operations. Thus the Argyll field may, because of the method of production adopted, be unable to achieve the production rates we have suggested owing to 'down-time' caused by bad weather. Again, the life of the Auk field may be significantly shorter than that shown in Appendix 2.1,[21] and we should bear in mind the fact that both fields have reserves in the same Zechstein dolomite formation as the Lockton gas field in Yorkshire, which was discovered in 1968 and subsequently developed only to be shut down for lack of reserves. Finally, the techniques applied are relatively untested. This is particularly relevant in the case of the Argyll field. We have assumed that the novel production method chosen for Argyll will prove to be successful, but experience of operating conditions may require some modification of this view and of our other assumptions.

Given these provisos, we believe that the resultant figures are a fair reflection of the variety of operating conditions that will be met in the North Sea. For the purposes of exposition we shall assume a 'ring fence' round each field so that each is treated separately for tax purposes.[22] We also apply the capital allowances currently available to the oil companies ('free depreciation') and we have priced output at $11 a barrel, which appears to be the market price that North Sea crude would realise at 1974 prices. Using this as the basis we can then calculate the likely profitability of the fields under different fiscal regimes.

The calculation of 'profitability' is not simply a matter of comparing the straight 'accounting' capital and operating costs with expected revenue, for the *timing* and profile of expenditure and revenue are also extremely important. As long as the rate of interest is positive, any individual or company will prefer a given income today to the same income at some future date.[23] For this reason a conventional calculation of profit based on accounting costs and returns will overstate the true profitability of any activity where capital expenditure is incurred ahead of the receipt of revenue. This is a very real problem in the North Sea, as almost all fields will require very heavy capital expenditure substantially in advance of production (e.g. Forties) and only a few will yield an income very quickly after the expenditure of relatively small capital sums (e.g. Argyll).

This problem is conventionally handled by applying discounted cash flow (DCF) techniques. This corrects for the 'time-value' of money by calculating the *present* worth or *present* value of a future stream of income. The specific application that concerns us here is the calculation of the *internal rate of return*, the rate of interest that reduces the present value of the future stream of income to zero. The internal rate of return then measures the highest rate of interest at which the project makes neither a loss nor a profit. Obviously, if the internal rate of return, so calculated, is greater than the company's cost of capital (equity or loan) then the project will be profitable, and vice-versa. For comparative purposes it may be useful to bear in mind the average internal rate of return on capital invested in British industry, which appears to be 15 per cent.[24]

We have calculated conjectural internal rates of return (IRR) on Argyll, Auk, Piper and Forties under four different financial regimes (see table 2.4):

1. royalties ($12\frac{1}{2}$ per cent), plus a corporation tax at 52 per cent (the initial system);
2. royalties ($12\frac{1}{2}$ per cent), plus a corporation tax and excess profits tax at 75 per cent (which appears to have been the system which the last Conservative administration would have implemented);
3. royalties ($12\frac{1}{2}$ per cent), a barrelage tax set at $4 a barrel, and corporation tax at 52 per cent;
4. royalties ($12\frac{1}{2}$ per cent), a barrelage tax set at $4 a barrel, and corporation tax and excess profits tax at 75 per cent.

The results show why the initial financial regime had to be abandoned: it would have yielded a very high rate of return on each field, particularly on Argyll. The expected returns were far greater

TABLE 2.4 ESTIMATED INTERNAL RATE OF RETURN: ARGYLL, AUK, PIPER, AND FORTIES

Financial Regime	Argyll (%)	Auk (%)	Piper (%)	Forties (%)
1	279	85	52	33
2	193	59	39	24
3	147	73	37	22
4	109	56	29	15

than those necessary to induce the companies to undertake exploration and exploitation, and some alternative system had to be found. Yet it is clear that an efficient and practicable solution will not be easily established, for profitability varies very greatly from one field to the next.

Nor is it easy, on the basis of established principles, to forecast which fields are likely to prove most profitable. Most informed observers will be surprised to find that the two largest fields, Forties and Piper, may be the *least* profitable on the IRR criterion. On a conventional accounting basis, dividing all capital and operating costs by total output over the life of the field, we would find a cost per barrel of $2.26 for Argyll, $2.74 for Auk, $1.12 for Piper and $1.88 for Forties, but the two smaller fields with the highest accounting costs are, on our calculations, more profitable than the two larger fields with the lower accounting costs. The apparent paradox is resolved if we recall an earlier discussion of the *timing* of expenditures and receipts. Argyll and Auk are more profitable simply because they will both yield a quick return on the capital invested, while both Piper and Forties require heavy capital expenditure some considerable time in advance of the receipt of income.

This complexity imposes substantial problems for the formulation of a taxation regime which should attempt to 'cream off' excess profits while not discouraging exploration and production. Obviously the regime suitable for one field may not be suitable for the next and each of the regimes whose effects are calculated above would leave more than one field with a very high internal rate of return.

Nonetheless, we may be able to arrive at some general conclusions. First, expected profitability varies so substantially between fields that it would be inefficient to collect taxes through any flat-rate output tax such as a barrelage tax. Depending on the flat rate applied a barrelage tax would: (*a*) leave some fields with substantial excess profits, (*b*) make some fields unprofitable, or, and this is the most likely, (*c*) result in some combination of (*a*) *and* (*b*). Moreover, a barrelage tax cannot take any account of variations in the market price of crude, which is of crucial importance in determining profitability. For similar reasons, the system applied by Arab countries is inappropriate to North Sea conditions; for a regime of posted, tax-reference prices, on which taxes are levied at a standard rate, is in effect a barrelage tax owing to the extremely small variation in cost conditions faced by each Arab producer.[25]

It follows from this that the taxation regime required must be sufficiently flexible to allow for varying field conditions. As we have seen (regimes 2 and 4), even a profits tax levied at 75 per cent would not contain sufficient flexibility to deal with the extremely profitable fields. The only alternative, so long as we rely on the taxation system alone, would be to combine an excess profits tax with a sliding scale of royalties, varying the royalty rate according to the profitability of the individual field. This has ample precedents; e.g. the Norwegians have adopted a sliding scale of between 8 and 16 per cent.

We have left to the last the system proposed by the present Labour administration, which is to combine measures we have already discussed, royalties (at the existing rate) and an excess profits tax, with government equity participation in the oil fields. Participation is of course a feature of the Norwegian system, and there is no doubt that the Norwegian example has been very influential in shaping Labour's policies. Yet it is not at all clear that the Norwegian example is relevant to the UK's position. Participation would give the State a direct share of profits, but it would be necessary to raise large capital sums to make this possible. It is doubtful that this would be beneficial in present circumstances. A substantial part of North Sea developments has been financed by direct inward investment from abroad,[26] and it would seem sensible to encourage this capital inflow given the UK's weak balance of payments position. Participation would require compensation and hence an outward flow of capital in the short and medium term.

Nor would participation solve the problem of which taxation regime should be adopted, as it would still be necessary to establish a fair and reasonable return on private equity capital (and for that matter a formula to determine adequate compensation for that part of equity capital brought under state ownership). Any system implemented could be used to ensure reasonable, but not excessive profits, without any recourse to equity participation by the State.

The essential reason for participation, which lies behind Norwegian policy, is that it allows greater control over the rate of oil production and provides the route through which a state oil company can become *actively* involved in oil exploration, production and the range of downstream processing and distributive activities. Hence, participation is tied to the idea that a British National Oil Corporation should be established[27] as a faithful carbon copy of Norway's Statoil. In Norway's case such a policy is perfectly logical, but, unlike Norway,

Britain has a number of companies with an established expertise in the industry – British Petroleum, Shell and Burmah Oil – and does not therefore have the same need to 'break into' the industry. Ultimately, of course, this argument comes down almost to an article of faith. The British National Oil Corporation can be regarded as a dynamic source of state enterprise which will establish a new force in off-shore technology. We must plead scepticism. We believe that it would be extremely difficult to recruit the personnel to undertake such an active policy and there is nothing about British policy in the North Sea, nor in the past performance of the public sector, which encourages us to pin much faith on the expertise, proficiency and dynamism of a state-controlled agency.

In sum, we consider that the activities and profits of the oil companies can be controlled by a sensible use of the fiscal weapons open to the government. For these purposes direct state control is unnecessary. We would stress that the taxation regime adopted must be designed to meet the immense range in the profitability of different fields. Indeed, as profitability varies so substantially, and is so difficult to predict from general principles, two further conclusions follow. First, it seems sensible to treat each field separately for taxation purposes.[28] Second, it will be very difficult to establish in advance the type of taxation system that will yield a reasonable, but not excessive, rate of profit on each field. It may well be preferable to determine the effective rate of tax *ex post* rather than *ex ante*, so that calculation rather than hunch and judgement would be the basis of action. This could be done, for example, by establishing whatever is considered to be a reasonable internal rate of return and, after full disclosure of costs and operating conditions, ensuring that this return is met on each field.[29] Alternatively, one could approach such a solution through a combination of an excess profits tax (say at 75 per cent) and a sliding scale of royalties.

2.5   What Rate of Production?

The key question here is whether the rate of production of North Sea oil should be determined by private or national interests. It seems to be generally assumed that the State would wish to reduce the rate of production substantially below that rate set by private interests.

Indeed, each major political party has indicated that production would be set close to that needed to offset crude oil imports and, on our calculations, this would require significant control of production by the late 1970s (see p. 95). But is such a policy desirable, and on what principles should one determine the optimum rate of exploitation?

Let us begin with the oil companies. The optimum rate of exploitation will be determined by:

1. the expected rate of interest;
2. the expected change of oil prices; and
3. the variation in costs associated with different rates of production.

It is often said that if expected future prices are stable then this justifies a slow rate of exploitation. This is incorrect. As long as prices are stable or declining an oil company will always follow a policy of quick exploitation provided the rate of interest is positive. This is so, because the oil company can then maximise its income over any time period by exploiting now and then reinvesting its profits. The company will defer production only if the expected rate of increase of prices is *greater* than the rate of interest, or if the costs involved in obtaining a more rapid rate of production more than offset additional revenues from that extra output. If, as we have suggested,[30] oil prices are more likely to fall than rise, and given that within quite wide ranges output can be varied without a very large variation in costs, it is highly likely that the oil companies would follow a policy of quick exploitation.

Contrary to the popular viewpoint, much the same considerations arise when considering the optimum depletion policy for the State. Government revenue will vary directly with private revenue and a quick rate of exploitation will be favoured if the expected rate of interest is above the expected rate of increase of oil prices. Therefore, if the State and the oil companies take the same view of these factors they are likely to follow a similar depletion policy. It is often argued that the depletion policy followed by the State and by private industry will differ because the State is less myopic and values future income more highly; i.e., the social rate of discount is lower than the private rate of discount. If this is so, then public and private interests may diverge.[31] However, the proposition that the State is less myopic is, in view of recent history, very difficult to accept.[32] The private and social rate of discount may well be very similar, and if the State wishes to maximise income in the future it will still be advised to encourage production if the expected rate of interest is greater than the expected increase in oil prices, for by taking the revenue now and investing it in interest-bearing assets (say, by exporting capital) it will maximise income at any given future date. If oil prices are stable and the social rate of discount is positive, then the rate of discount will determine not the rate of production, but the extent to which oil revenues are consumed now or reinvested in capital assets!

Given this, and our view that oil prices may weaken somewhat in the future, it may be rational for the State to exploit North Sea oil reserves quickly. Certainly a crude rule of thumb of relating output to import needs is a very inadequate guide to production policy. On the other hand, there are some factors that do suggest that the State might follow a slower rate of exploitation than the oil companies would dictate. The State is likely to put considerable weight on strategic factors, on the need to establish a British expertise in off-shore technology, which may be encouraged by a slower and steadier buildup of production, and on the social and political costs resulting from the substantial stresses which rapid development might place on the communities directly affected. None of these considerations would weigh as heavily with private interests. Some state intervention in determining the rate of production is therefore to be expected.[33] It should, however, be accompanied by an explicit recognition that taxation and production decisions are interrelated. They cannot be implemented separately for, as we have seen, under any taxation system the rate of return is not independent of the timing of revenues, which depend on the rate of production.

2.6 CONCLUSIONS

We have emphasised that the owner of property rights in petroleum can enjoy, and does certainly enjoy in present and likely future circumstances, a large monopoly rent. Hence, when in selling production licences the State transfers exploitation rights to commercial interests, it should attempt to ensure that it captures these rents while leaving the commercial interests with sufficient incentive to continue exploration and production.

At the present time the minimum aim of any new financial regime must be to capture the windfall gains and added economic rent that would otherwise accrue to oil companies as a consequence of the large increases in oil prices that have occurred since the fourth licensing round in 1971–2. However, whatever system is adopted, we believe it should offer the commercial interests involved in North Sea exploration a higher rate of return on the capital than the 'normal' rate of return for the economy as a whole. This would reflect some margin for exploration costs which are not readily attributable to particular fields[34] and a risk premium reflecting the greater uncertainty which attached to the exploration and exploitation of natural resources. Most authorities accept that an element of greater risk does arise although it is occasionally disputed.[35] Investment in natural resources involves special difficulties which are most apparent in exploration for oil and gas. The would-be producer has to face the possibility that he will invest substantial capital sums without obtaining *any* final marketable product, a prospect which is extremely unlikely in manufacturing industry, but which is more marked in petroleum exploration than in exploration for other materials.[36] This risk must become greater in the future as exploration moves to less promising territory and the dry-hole ratio increases.

In the UK sector of the North Sea these difficulties, which are always attendant on exploration, were magnified by the nature of the licensed territory – deep, unknown and stormy water, requiring new and untried operating techniques. It is not surprising that a number of companies have invested large capital sums without finding any significant petroleum deposits. Further, although oil companies may be risk-takers rather than risk-averters by nature, the number of risk-takers tends to decline when the bookmaker reserves the right to change the starting prices after the race has been run! This is precisely what we have suggested in advocating a regime that varies the taxation system *ex post* to 'cream off' excess profits. There may be disagreement over what constitutes an adequate target internal rate of return, but something in the region of twenty-five per cent may be appropriate.[37] Certainly, given the UK's present balance of payments situation, it would seem sensible to err slightly on the side of generosity rather than parsimony. The motto of taxation policy should be *dignue est operarius mercede sus*.

ESTIMATED CAPITAL EXPENDITURES, OPERATING COSTS AND OUTPUT FROM ARGYLL, AUK, PIPER AND FORTIES FIELDS[1]

	Argyll			Auk		
	Capital expenditure ($m)	Operating costs ($m)	Output ('000 barrels/day)	Capital expenditure ($m)	Operating costs ($m)	Output ('000 barrels/day)
1974	25	10	10	45	—	—
1975		40	30	45	15	20
1976		40	60		25	40
1977		40	60		25	40
1978		40	60		25	40
1979		40	60		25	35
1980		40	50		25	35
1981		40	50		25	35
1982		40	50		25	35

	Piper			Forties		
1972	100	—	—	150	—	—
1973	200	—	—	300	—	—
1974	200	10	20	600	—	—
1975		30	100	300	30	50
1976		30	225	150	75	200
1977		30	225		75	400
1978		30	225		75	400
1979		30	225		75	400
1980		30	202		75	360
1981		30	182		75	324
1982		30	164		75	292
1983		30	148		75	262
1984		30	135		75	236
1985		30	121		75	213
1986		30	109		75	191
1987		30	98		75	172
1988		30	88		75	155
1989		30	80		75	139
1990		—	—		75	126
1991		—	—		75	113
1992		—	—		75	102
1993		—	—		75	92
1994		—	—		75	82

¹Based on data supplied by Wood, Mackenzie and Co.

NOTES TO CHAPTER 2

1. Appropriately enough, this was the text of the sermon at the launching of Highland One from Highlands Fabricators' yard at Nigg Bay in the Cromarty Firth.

2. This does not of course confer sovereignty over the high seas.

3. Minister for Power *Hansard (Commons)* (23 March–10 April 1963–4).

4. In fact, two types of licences are involved – production and exploration licences. The text above describes the former, as these are the more important.

5. The licensee has an option to continue the licence for the remaining half of his area for a further forty years.

6. Minister of Power *Hansard (Commons)* (12 July–23 July 1964–5).

7. This compares with the position at April 1974 when it was estimated that the UK share was 40 per cent, the US share 37 per cent and others 22 per cent. See *United Kingdom Offshore Oil and Gas* Edinburgh, Scottish Council (Development and Industry) (1974).

8. The Celtic Sea was a new term coined to distinguish the areas licensed from the Irish Sea, as this stretch of water was formerly known. It then attempted clearly to establish UK ownership, possibly at the cost of encouraging Welsh proprietary claims!

9. G. L. Reid, K. Allen and D. J. Harris *The Nationalised Fuel Industries* London, Heinemann (1973) p. 118.

10. The number of exploration wells drilled fell over 1969–71 (see table 3.1, p. 58), while drilling activity increased slightly in the Norwegian and Netherland North Sea sectors.

11. First Report from the Committee of Public Accounts, *North Sea Oil and Gas*, 1972–73, paras 2778–3021.

12. See T. Hayward 'No More Drilling until Gas Prices Rise' *Offshore Services* (January 1974).

13. The government take in Middle East countries is now probably in the region of 95 per cent and the Norwegian government take is at the top end of the quoted range.

14. Most notably by Lord Balogh. See *Hansard (Lords)* (24 May 1971, 2 December 1971, 7 June 1972 and 13 December 1972).

15. David Ricardo *The Principles of Political Economy and Taxation* Cambridge, Cambridge UP (1951) vol. I, p. 67.

16. See p. 65

17. Outright nationalisation would also mean that *all* the risk was borne by the State.

18. A further argument against nationalisation is that it would put at risk the overseas investments of the UK oil majors.

19. See K. W. Dam 'Oil and Gas Licensing and the North Sea' *Journal of Law and Economics* (October 1965).

20. The estimates are those published by Wood, Mackenzie and Co., members of the Stock Exchange, Edinburgh. The upward adjustments are based on discussions with Wood, Mackenzie and Co.

21. An alternative production profile to the one shown for Auk in Appendix 2.1 would be 10,000 barrels per day in 1975, 40,000 b/d in 1976–7, 35,000 b/d in 1978 and 10,000 b/d in 1979.

22. As we shall see, such an assumption may become a reality.

23. For the simple reason that income earned now can then be invested and will earn a return given by the rate of interest.

24. J. M. Samuels and D. J. Smyth 'Variability of Profits and Firm Size' *Economica* (1968). The current rate of return is certainly lower than this, but hardly reflects a suitable standard of reference.

25. See p. 8.

26. See p. 103.

27. See Lord Balogh 'Government and the North Sea' *The Banker* (September 1974).

28. In other words, we should have a ring fence round each field and not simply round North Sea operations as a whole, as is currently proposed.

29. Such a system would require some state oversight of oil company expenditure, for in the absence of this there would be little incentive to adopt economically efficient techniques of production.

30. See p. 14.

31. Suppose, for example, that the expected annual rate of increase of oil prices is 7 per cent, that the private rate of discount is 8 per cent and the public rate of discount is 6 per cent. Then the oil companies will wish to exploit, while the State will wish to conserve, reserves.

32. It is widely accepted in welfare economics, for reasons that are obscure. The State's decisions are determined largely by politicians and it is not obvious that their planning horizon is longer than that of private companies or private individuals.

33. Indeed, it is already foreshadowed by government action. See the *Financial Times* (20 September 1974).

34. In estimating internal rates of return we have made no allowance for exploration costs. This is not a serious error, for these costs (see p. 75) are not large relative to capital and operating costs; but obviously some allowance for exploration costs should be made in constructing an appropriate taxation regime.

35. See M. V. Posner *Fuel Policy: A Study in Applied Economics* London, Macmillan (1973) pp. 180–3 and 204–9.

36. See H. Steele 'Natural Resource Taxation: Resource Allocation and Distribution Implications' in M. Gaffney *Extractive Resources and Taxation* London, University of Wisconsin Press (1968) p. 241.

37. Some oil executives have suggested that a lower internal rate of return of 15 per cent would be satisfactory, while others have suggested 25 per cent as a 'minimum'. See *The Guardian* (27 July 1974) and *The Sunday Times* (22 September 1974) p. 57. Lord Balogh appears to believe that 25 per cent would be a reasonable rate of return. See *Hansard (Lords)* (13 December 1972) col. 710.

CHAPTER 3

A Maid of Norway's Legacy[1]

When the time came for the princess to leave her home, however, Christian found himself in straits. Of the 10,000 florins he had undertaken to pay on her departure he could muster only 2,000; and as a pledge for the remainder he was forced to give up the Shetland Islands, which, unredeemed like the Orkneys, have remained a Scottish possession from that day to this. (P. Hume Brown, *History of Scotland* vol. 1, p. 210)

3.1 INTRODUCTION

Until 1469 the Orkney and Shetland Islands were part of the Kingdom of Denmark, Norway and Sweden, which at that time were ruled by King Christian. In 1468 a treaty of marriage was signed between James III of Scotland and Margaret, daughter of King Christian. As his daughter's dowry Christian was to give 60,000 florins, of which 10,000 were to be paid immediately and the Orkney Islands given in pledge for the rest. In the event the Shetland Islands were also included in the dowry. Events five hundred years later have added a great deal of irony to this transaction because most of the oil and gas discoveries in the North Sea have been in waters lying immediately to the east of Orkney and Shetland. Fortune has favoured the Scots – apart from the 1707 Treaty of Union!

The purpose of this chapter is to provide a description of the physical considerations surrounding the North Sea discoveries and of how events have developed over time. Above all, it is concerned with assessing the extent of the oil and gas reserves which are likely to be exploited and the possible buildup of production. The reader familiar with the estimates produced by various authorities will be aware of the many difficulties and dangers inherent in this process. At times it does appear, as has often been claimed about economists, that there are as many estimates as there are experts!

Nevertheless, our subsequent analysis must be based on some view of the likely size of reserves and we therefore attempt in this chapter to establish the broad outline of the problem and to narrow down, as far as is possible, the area of uncertainty. We believe that the estimates produced are the best that can be made given present knowledge and they will not be badly falsified · by events. Yet, inevitably, the estimates must be subject to some margin of error and we have therefore set out the basis on which we have proceeded so that the interested reader can appreciate the complexity of the problem.

Hence, the chapter falls into three parts. Firstly, we look at the complex geology of the North Sea. Secondly, we explain briefly the technical issues involved in exploration and production. Thirdly, we assess the extent of established oil and gas fields and comment on future prospects. Throughout the discussion we have tried to present the main issues and facts in a manner comprehensible to the layman; for some understanding of these matters, however elementary, is necessary for the elaboration of our subsequent arguments.

3.2 The Geology of the North Sea and its Petroleum Resources

Care is required in the use of terminology, as in some instances there are significant differences between the popular and technical meanings of various words. Briefly, all the discoveries in the North Sea have been of types of petroleum (which is distinct from petrol, which is a product obtained from petroleum). Petroleum occurs naturally in rocks and is a mixture of hydrocarbon compounds, sometimes with impurities such as nitrogen and sulphur. Essentially, petroleum can exist in three forms – as a solid, a liquid or a gas – and there are important differences between these three forms. In its solid form petroleum is made up of the heavier hydrocarbon compounds and occurs naturally as asphalt, tar, bitumen, etc. These are often found on the surface of the earth, as in the famous pitch lake in Trinidad. Petroleum in its liquid form is known as crude oil and is usually beneath the earth's surface, although on occasion it is found on the surface as seepages. Liquid petroleum is the feedstock for the refineries that produce petrol, fuel oil, diesel oil etc. Petroleum in its gaseous form is known as natural gas (as distinct from manufactured or town gas) and comprises the lightest hydrocarbon compounds. Again this occurs underground but occasionally escapes to the surface. It has been suggested, for example, that the fiery furnace of Shadrach, Meshach and Abednego in the Old Testament was a natural gas seepage which had been ignited.[2]

The crucial factors determining the form in which petroleum exists are pressure and temperature. Each form can be converted to the others by altering the pressure and/or temperature. In the North Sea the discoveries have been of liquid and gaseous petroleum. Natural gas has been found on its own – chiefly in the southern parts of the North Sea – but crude oil is invariably found in association with gas. The main reason for this is that the transference of oil from underground reservoirs to the surface involves a reduction of pressure which will usually convert some of the oil into gas. This associated gas may or may not be produced in commercially exploitable quantities; if not it is usually reinjected into the reservoir or burned off from flare stacks.

For petroleum to occur, there must obviously be suitable geological conditions. The first of these is the existence of appropriate source rocks, which for petroleum are generally believed to be sedimentary, i.e. those made up by the deposition on the sea floor of marine organisms such as plankton and bacteria. Over time this sediment became buried as more material was deposited from the rivers, and as the depth of burial increased, pressure and temperature resulted in the formation within the sedimentary rock of hydrocarbon molecules. The North Sea is a sedimentary basin of this type and the appropriate source rocks were laid down during the Carboniferous, Jurassic and Tertiary periods – between 350 million and 2 million years ago. During and since these periods the geological history of the North Sea has produced strata which in some cases are conducive to the formation of petroleum.

There are certain crucial conditions. There must exist above the source rock suitable strata of reservoir rock and cap rock. The reservoir rock is that in which the various forms of petroleum accumulate and the rock has therefore to be porous and permeable to allow the hydrocarbon molecules to move vertically or laterally through the salt water trapped in the pore spaces. The molecules will continue to move upwards until they encounter impermeable rock through which they cannot move. If the latter rock exists, then a pool or reservoir of petroleum will build up beneath this cap rock. Various geological structures such as faults, anticlines and stratigraphic traps are particularly useful in the accumulation of petroleum deposits.

Rather simplistically, then, petroleum deposits can occur where there have existed over time suitable combinations of source rock, reservoir rock and cap rock. The necessary conditions are so severe that deposits will be infrequent and often difficult to detect. Oil and gas exploration is simply the search for such deposits. Although the brief description above may imply that the formation of petroleum is straightforward and common, this is far from being the case and it should be remembered that in the case of natural gas the process has taken up to 350 million years to reach its present stage.

The North Sea is a vast sedimentary basin laid down millions of years ago and in fact many parts of western Europe presently above the sea level are part of the same basin. The basin is far from being uniform and within it it is possible to identify a series of subsidiary basins which are being successively explored. To date the oil and gas fields have been discovered in three main areas – the southern North Sea basin, the middle North Sea basin and the East Shetland basin.[3] The frontispiece illustrates the geographical locations of the fields. These subsidiary basins display significantly different geological characteristics, and one very encouraging feature of the discoveries has been that they have been made in a range of reservoir rock strata. For example, the discoveries in the southern North Sea have been virtually all of natural gas, mainly because of the relatively low depth of the source rocks which are believed to be coal seams of the Carboniferous period. In contrast, the oil discoveries further north originate from rocks of the Mesozoic and Tertiary eras. Generally, the crude oil that has been found has been of the lighter variety, making it suitable for refining into, *inter alia,* motor spirit. Discoveries of the heavier crude oils have been small, and for the United Kingdom this implies that some of North Sea production will be exported while the UK will continue to import heavier crudes from the Middle East and elsewhere.

3.3   THE TECHNICAL ISSUES OF EXPLORATION AND DEVELOPMENT

The point was made earlier that, although the source of virtually all petroleum deposits is sedimentary rock, deposits can occur both under dry land and under water. To a large extent the technology required to extract and produce oil and gas is the same on-shore and off-shore. Off-shore discoveries, however, involve the additional problems of water and weather and consequently make exploration and development more difficult and much more costly.[4]

Although off-shore exploration has a fairly lengthy history, it has mainly taken place in shallow in-shore waters and the North Sea has presented an enormous challenge which has virtually necessitated the devising of a new technology. In the North Sea equipment has to be capable of coping with the legendary storm that occurs once every hundred years – in other words, the one-hundred-foot wave – and there are additional problems arising from the variable nature of the seabed which can consist of mud, quicksand, sand waves, exposed rock, etc. As similarly inhospitable areas, such as the Arctic Ocean and the Bering Sea, are attracting oilmen, the pressure for technological advance is one reason why the North Sea attracts so much attention. If successful, the production methods used in the North Sea will undoubtedly be used in these other areas.

For ease of exposition it is useful to identify three distinct phases of activity; exploration, manufacturing and production. Of necessity these three phases overlap but they do display different characteristics. Taking the exploration phase first, in any area the first stage is to locate geological formations that may hold petroleum deposits. Various techniques can be used, of which seismic surveying is the most common, and the results obtained are analysed in conjunction with available geological information in order to determine precisely where drilling should take place.

Exploration drilling is undertaken by rigs mounted on drilling installations. The first mobile off-shore drilling rig was used off Louisiana in 1949; at the present time there are about three hundred in operation with another hundred under construction or on order. There are basically three types of rig in use and the choice of type usually depends on water depth and expected weather conditions. The first type is known as the jack-up rig, which is bottom-supported in that the platform is jacked up on legs which rest on the seabed. To move the rig the legs are raised from the seabed. This type of rig is most suitable for calm and shallow waters (no more than 200 feet) and has been extensively used in the Gulf of Mexico and the southern North Sea. Unfortunately it has not proved suitable for the northern North Sea and is only occasionally used during the summer months. The second type is the drill-ship, which is essentially a ship that has been converted or specially built to allow drilling to take place through the hull or over the side. Drill-ships offer certain advantages, particularly in mobility, over jack-up rigs but again they are not well suited to the weather conditions in the northern North Sea, although they are being increasingly used for deep-water drilling elsewhere. The third and most popular type is the semi-submersible rig of which there are quite a few variants. The basic principle of operation is the existence underwater of large pontoons which support the drilling platform above water and which offer vastly increased stability. Both construction costs and operating costs are much higher than for jack-up rigs but larger and more advanced semi-submersibles are being increasingly used in the North Sea. These rigs can be self-propelled or can rely on tugs to take them from location to location.

Whichever type of rig is used, there is a need for back-up facilities to provide and transport the personnel, equipment and materials required to drill an exploration well. This involves the establishment of an operating base, a harbour base for sea transport of materials and equipment and a helicopter base for the transport of personnel and certain urgent items. There are great advantages in having these bases together if possible and they are located on-shore as near as is possible to where exploration activity is occurring: hence the establishment of bases on the east coast of Scotland in Aberdeen (principally), Peterhead, Lerwick, Dundee etc. and, further south, in Great Yarmouth. On average a rig requires two supply boats to keep it supplied with cement, drilling fluids, casing etc., and contracted access to helicopter services. In the supply bases that have been established there has built up a large number of specialist companies. Only rarely do the oil companies own and operate their own exploration rigs and this is usually done by specialist companies, as are transport operations etc. The scale of activity can perhaps be understood by remembering that the latest version of semi-submersible rig costs about £25 million to build with a daily operating cost to the oil company of up to £30,000.

If oil or gas is discovered then additional facilities are required, which can be grouped under three headings:

1. a fixed production platform (or platforms) with ancillary equipment to gather the hydrocarbon deposits;

2. an underwater pipeline or tanker-loading facilities to transport the oil or gas to shore;

3. an on-shore terminal.

The fixed production platform is similar in many respects to the exploration rig. It is used to drill the development (production) wells, which may number up to forty, and it supports the wellhead equipment, gathering and treatment plant, power generators and personnel accommodation, which are built separately, usually in modular form. Often there is a series of platforms. Until recently, most production platforms were steel structures piled into the seabed, their height and size depending on the water depth and the nature of the production equipment required. In the last two years, however, gravity structures have become popular for water depths exceeding 350 feet because of their storage capacity and the elimination of the need for piling, which is a difficult and expensive task in the North Sea. Such structures could be of steel or concrete but concrete is usually preferred. There are quite a few fields in the North Sea, however, where the seabed conditions have been such as to necessitate the use of a piled platform. A possible alternative to fixed platforms is sub-sea completion, with the wellhead and production equipment located in the seabed, but the technology is still at an early stage and it is unlikely that this method will be used on a large scale in the North Sea until the 1980s. Finally, mention should be made of the Argyll field which, because of its short lifespan, is being developed with the use of a converted exploration rig anchored to the seabed.

Turning to the transportation system, there are two alternatives: a pipeline system to transport the oil to shore, or a tanker-loading facility which allows the oil to be loaded at sea into tankers. The latter is often used for small fields where output is insufficient to justify the cost of a pipeline (as with the Auk and Argyll fields in the North Sea) or as a temporary measure while a pipeline is being fabricated and laid (as with the Ekofisk field). Gas has to be piped ashore although associated gas is frequently reinjected if it is not present in commercial quantities.

When the oil or gas reaches shore it has to be treated and/or refined, although this does not necessarily have to be done in Scotland or the UK. A large proportion of North Sea oil will be transferred to tankers and exported in crude form to refineries elsewhere. With gas this is much more difficult and costly as it involves liquefaction and gas is usually transferred into an off-shore pipeline network. Frequently, petro-chemical industries using oil or gas or petroleum products as feedstocks are located close to the refineries and landfall terminals.

The remaining phase identified above was the manufacturing phase, which has been subsumed largely in the brief descriptions of the exploration and production phases. There are certain features, however, that merit separate mention. The manufacturing phase involves the fabrication or manufacture of the range of equipment and materials required. Some of these facilities (exploration rigs, supply boats etc.) are mobile by definition and therefore can be constructed anywhere in the world and moved to the North Sea. On the other hand, items such as production platforms are usually built close to the fields, as are the landfall terminals.

3.4 RESERVES AND PRODUCTION

The above descriptions of the physical and technical issues are intended to provide a background to the remaining part of this chapter and to the discussions of on-shore economic and financial aspects in later chapters. At this juncture it may be useful, therefore, to look at the scale of discoveries in the North Sea and at how they fit in with the industry worldwide. We shall attempt to assess the extent of the oil and gas reserves in the UK sector that have been established by exploration up to 1974 and of the reserves that may be discovered in the future. Finally, we shall consider in detail the likely rate of oil production to 1983 and, in more general terms, subsequent production levels.

The first on-shore wells were drilled in Lake Maracaibo, Venezuela, in the early 1920s. In open waters the first wells were drilled in the Gulf of Mexico in the 1940s. It was not until the early 1960s, however, that exploration began in the North Sea. The precipitant was the on-shore discovery in 1959 of natural gas at Slochteren in the Groningen province of the Netherlands. It seemed certain that structures similar to the Slochteren field extended off-shore, possibly across to the English coast where there had been some small on-shore discoveries of gas; and hence there arose great interest in undertaking exploration work in the southern North Sea. Apart from seismic surveying, this had to await agreement on the territorial rights of the countries surrounding the North Sea. Sufficient agreement was reached on the basis of the 1958 Geneva Convention and following its ratification by the United Kingdom in 1964 exploration drilling began. Table 3.1 shows the pattern of activity since that date.

The first discovery of hydrocarbons in the North Sea was the West Sole gas field which was discovered by BP in 1965. The West Sole field is about forty miles from the Humber estuary. Production started in 1967 by way of a pipeline to Easington at the mouth of the estuary. In 1966 three more large gas fields (Hewett, Leman Bank and Indefatigable) were discovered off the Norfolk coast and the next few years brought additional finds in the southern North Sea, including the Dutch sector. A list of the fields, together with estimates of reserves and production rates, is given in table 3.2. As all these fields with the exception of Rough and Frigg are in production the margin of error for these estimates is likely to be small.

TABLE 3.1 EXPLORATION WELLS DRILLED

(a) *UK sector of North Sea*

	East Shetlands	East Scotland	Southern	Total
1964			1	1
1965			10	10
1966			20	20
1967		7	35	42
1968		1	30	31
1969		8	34	42
1970		10	12	22
1971	4	13	7	24
1972	9	16	8	33
1973	16	18	7	41
Totals	29	73	164	266

(b) *all North Sea sectors*

Sector	Number, 1964–73
UK	266
Norway	80
Denmark	20
West Germany	90
Netherlands	12
	468

TABLE 3.2 ESTIMATED RESERVES OF KNOWN COMMERCIAL GAS FIELDS

Discovery	Production	Field	Block(s)	Licences	Reserves (10^{12} cu. ft)
1965	1967	West Sole	48/6	BP	1.0
1966	1969	Leman Bank	49/26	Shell/Esso	14.0
			49/27	GC/Amoco group	
			49/28	Arpet/Sun Oil group	
			53/2	Mobil	
1966	1971	Indefatigable	49/23	GC/Amoco group	8.0
			49/19		
			49/24	Shell/Esso	
1966	1969	Hewett	48/29	Arpet/Sun Oil group	3.5
			48/30	Phillips/Fina group	
			52/5a	Phillips/Fina group	
1968	1972	Viking	49/17	Conoco/NCB	5.0
			49/12a	Conoco/NCB	
1968	1975	Rough	47/8	GC/Amoco	1.0
			47/3a	GC/Amoco	
1972	1976	Frigg UK	10/1	Total Group	7.0
1971	1978	Frigg Norway	25/1	Petronord	8.0

Although these gas discoveries were significant in the context of UK energy requirements, they were not outstanding and exploration activity began to lag as can be seen from table 3.1. Most companies were hoping to find oil rather than gas. The first oil discovery was made in fact in the Danish sector in 1966 but this was a very small non-commercial find. A major stimulus to and redirection of effort, however, came in December 1969 with the discovery in Norwegian waters of the large Ekofisk oil field. This is located just north of the 56°N parellel and sparked off the massive scale of activity which we have seen in the northern North Sea in recent years.

In the UK sector the first oil find was made by the Gas Corporation/Amoco group with their Montrose field, which was discovered in December 1969 but not announced as commercially exploitable until four years later. In the meantime BP discovered the Forties field in November 1970 and announced its commercial viability in December 1971. To date[5] there have been eighteen commercial finds in the UK sector; other finds have yet to be fully appraised and there is still a considerable amount of exploration drilling to be undertaken.

All the oil fields have been discovered in the northern North Sea – in Scottish waters – where the geology is different from that in the southern North Sea basin; and the fields can be arranged roughly into three groups:

1. the southern group, just over the boundary from Ekofisk, and including Argyll and Auk (the area extending from 56°N to 57°N);

2. the Aberdeen-Orkney group, which includes Forties, Piper, Beryl and the Frigg gas fields (57°N to 60°N);

3. the East Shetland group, which includes Brent, Thistle and Ninian (60°N to 62°N).

Of these areas the East Shetland basin has proved to be by far the most rewarding and has been the scene of the most intensive drilling activity.

It will be seen from table 3.1 that most of the exploration drilling has taken place in UK waters and this is both a cause and a consequence of the fact that most of the large finds have been in the UK sector. In the Danish sector only the Dan oil field has been declared commercial, although there have been a few smaller finds. Some large gas deposits have been found in the Dutch sector and an oil find was announced in April 1974 but has yet to be appraised. The West Germans have met with even less success, as the only gas deposits discovered are unsuitable for exploitation because of their very high nitrogen content. The Norwegians have adopted much stricter controls over the rates of exploration and activity has been on a small scale compared with that in the UK sector: discoveries of both oil and gas, nevertheless, have been encouraging and it is expected that those areas north of 60°N and adjacent to the median line with the UK will prove rewarding.

There is no doubt now that the oil and gas discoveries in the UK sector have been substantial and have exceeded most expectations. Our own estimates of recoverable reserves from known discoveries are given in table 3.3.

These estimates are based on information provided by the oil companies and, in the cases where such information was not obtainable, from oil industry journals. We believe that they are the best we can make with the information currently available and do not think that they will be proved wildly wrong when additional information becomes available. They are about 25 per cent higher than the official estimates produced by the Deparment of Energy[6] but they are significantly below the estimates of other observers.[7] For this

TABLE 3.3 ESTIMATED RECOVERABLE RESERVES OF KNOWN
COMMERCIAL OIL FIELDS

Field	Block(s)	Licensee	Estimated reserves (million barrels)
Alwyn	3/14	Total/Elf group	300
Argyll	30/24	Hamilton group	150
Auk	30/16	Shell/Esso	100
Beryl	9/13	Mobil group	500
Brent	211/29	Shell/Esso	2000
	3/4	Texaco	
Claymore	14/9	Occidental group	600
Cormorant	211/26	Shell/Esso	400
Dunlin	211/23	Shell/Esso	600
	211/24	Conoco/Gulf/NCB	
Forties	21/10	BP	1850
	22/6	Shell/Esso	
Hutton	211/28	Conoco/Gulf/NCB	500
	211/27	Amoco Group	
Maureen	16/29	Phillips Group	450
Montrose	22/18	GC/Amoco	200
	22/17	GC/Amoco	
Ninian	3/8	BP Ranger group	2100
	3/3	Burmah Group	
Piper	15/7	Occidental group	800
Thistle	211/18	Burmah group	750
Heather	2/5	Unocal group	350
Andrew	16/28	BP	400
Magnus	211/12	BP	750
Others*			1000
Total			13,800

*Discoveries in blocks 20/5 (Texaco), 21/1 (Transworld), 9/8 (Hamilton) and 9/13 (Mobil) for which no information is yet available

reason they may be considered cautious, but we think it important to distinguish between careful calculations based on actual data and speculation based largely on guesswork. Even the estimates in table 3.3, which are wholly based on known commercial finds, are subject to a fairly wide margin of error in individual instances. Obviously, reserves cannot be evaluated properly until a field is producing oil and there are many examples of past errors in estimates made in advance of actual production.

In table 3.3 it is estimated that at present known recoverable reserves of oil in the UK sector of the North Sea total approximately 13,800 million barrels (about 1800 million tons).[8] Three main factors will result in the revision (presumably upwards) of these figures:

1. the appreciation factor;
2. the recovery rate;
3. future discoveries.

Taking the appreciation factor first, reserve and production estimates are usually revised upwards through time. Firm figures cannot be provided until oil has been flowing for quite a few years and the reservoir characteristics are much better known, and consequently announcements at the time of discovery or after additional appraisal drilling are usually cautious. There is therefore an appreciation factor, which relates announcement figures to actual figures produced later, and this appreciation factor has attracted a great deal of attention. The main prophet is Odell, who maintains that for the North Sea the appreciation factor will be about two – in other words that the discoveries will prove to be twice as large as those originally announced.[9] Support for this theory comes from Albertan experience (with an appreciation factor of nine). What lessons this has for the North Sea is a matter of opinion. Seismic surveying and reservoir estimation have improved enormously in recent years and our own view is that Odell's estimates will prove to be too high but that reserves will be uprated generally over time – probably by not less than 25 per cent on average.

Secondly, it is quite possible that there will be improvements in the recovery rate. Conditions in the North Sea basins are such that the oil companies can recover only a proportion of the total reserves in the reservoirs – in most cases the proportion will be between 30 and 40 per cent. The high price of crude oil and technological advances, however, make it possible to increase recovery rates at economical costs. Already in the North Sea additional equipment is being installed in the Leman and Indefatigable gas fields to maintain delivery pressures in the face of declining reservoir pressures. As the oil reservoirs are run down it is likely that similar measures will be introduced to lengthen the lives of the fields and, given the present price of oil, it is possible that the companies will attempt eventually to add another 10–15 .per cent to recovery rates in order to achieve secondary and tertiary production. If such attempts were successful, then there would be a corresponding increase in recoverable reserves.

Thirdly, what of the future? It is obviously to be expected that future discoveries will add further to reserves. A large part of the North Sea has still to be drilled and in addition exploration activity has increased in the Celtic Sea and the West Shetland basin, for example. It is unlikely, however, that the present high rate of success will continue because the exploration companies are now drilling on the structures they regarded as most promising and will subsequently have to move on to less promising structures. In particular, the areas west of the British mainland do not appear to be as encouraging as the East Shetland basin; and the Celtic Sea is similar to the southern North Sea basin which suggests that discoveries there would more likely be of gas rather than oil.

Obviously, it is pure speculation to try to provide figures for the possible addition to reserves, but there are ways of reducing the areas of doubt. Firstly, it is possible to estimate fairly accurately the numbers of exploration rigs that will operate in the North Sea in the next few years. Secondly, on the basis of the estimated number of wells to be drilled, forecasts can be made using various success rates. Thirdly, forecasts can be based on the proportions of structures that have been tested and are still to be tested, using geological data to weight the possible success rates. Combining the information derived from these processes at least provides a foundation from which to speculate, and this has been attempted by various people, notably Dr J. Birks of BP Trading. In December 1973 Dr Birks estimated that at that time established recoverable reserves of oil totalled 11,500 million to 13,000 million barrels in the North Sea, of which about 8500 million to 9000 million barrels were in UK waters. The average success rate in the northern North Sea (north of 56°N and including Danish waters) for exploration wells was roughly 1:8, cf. 1:20 in the southern North Sea and 1:15 for the world. By the end of 1973 about 80 prime prospects had been tested out of a possible 220. Applying slightly lower success rates for future drilling, Dr Birks arrived at an estimate of total recoverable oil reserves in the North Sea of 38,000 million barrels.

Revising these calculations on the basis of the estimates of known recoverable reserves above and on more up-to-date geological data, particularly on prospects in the West Shetland basin and in Norwegian waters, produces an estimate of ultimate total reserves of around 40,000 million barrels, of which approximately 32,000 million barrels would be in UK (mainly Scottish) waters. The latter would amount to forty years' oil consumption at 1974 levels.

At least, then, it would appear that North Sea oil guarantees self-sufficiency for the UK for fifteen to twenty years (if consumption continued at the current level) and at best it might guarantee self-sufficiency for forty years. This is futurology, however, and for many purposes the rate of production over the next decade is of greater importance. In table 3.4 we have set out the expected production profiles for the fields listed in table 3.3.

TABLE 3.4 ESTIMATED PRODUCTION RATES FROM KNOWN OIL FIELDS
(THOUSANDS OF BARRELS PER DAY)

	1975	1976	1977	1978	1979	1980	1981	1982	1983
Alwyn			25	80	100	100	100	90	80
Argyll	30	60	60	60	60	50	50	50	—
Auk	20	40	40	40	35	35	35	35	—
Beryl		60	120	150	150	150	135	120	110
Brent		75	125	250	370	480	480	480	480
Claymore			25	60	150	150	150	150	135
Cormorant				40	75	75	75	68	60
Dunlin			30	100	125	125	125	125	110
Forties	50	200	400	400	400	360	324	292	262
Hutton			25	60	120	120	120	110	100
Maureen			25	60	100	100	100	100	90
Montrose		30	50	50	50	50	45	40	35
Ninian				50	200	350	500	500	500
Piper	20	100	225	225	225	225	202	182	164
Thistle			50	100	200	200	200	180	160
Heather				25	100	100	100	100	90
Andrew				25	100	100	100	100	90
Magnus				50	150	200	200	200	180
Others					50	200	400	400	400
Totals	120	565	1200	1825	2760	3170	3441	3322	3046

It will be seen that the normal phasing of a field development programme is a slow, three- to five-year buildup phase during which the production wells are drilled and brought into operation, followed by three years' peak production and a gradual rundown with output falling by approximately 10 per cent each year. Smaller fields like Auk and Argyll will usually have a quicker buildup, and if two drilling rigs can be installed on the production platform(s) development drilling can be speeded up.

We should make it clear that the achievement of starting dates and production levels set out in table 3.4 is contingent on the plans of the companies being realised. This may appear to be an heroic assumption in view of past experience, as to date no company has been able to meet its targets and 'slippages' have been frequent and substantial. Nevertheless, these delays have in part been due to over-optimism, which has presumably been dispersed; and more realistic time-scales are now being suggested. Moreover, bottlenecks are disappearing (e.g. regarding production platform sites) and we do feel that the estimates in table 3.4 can be achieved and maintained throughout the 1980s. We have adopted these as the central estimates for the remainder of the book.

Table 3.4 suggests that by 1980 the production of oil from the UK sector will have reached about 3 million barrels per day (150 million tons per year) and that this level will be maintained as later discoveries come onstream. On the basis of the earlier estimates of possible future discoveries we believe that a production rate of 3.6 million barrels a day (180 million tons per year) can be achieved by 1982 and sustained throughout the remainder of the decade.

As regards the production of natural gas from the North Sea, similar calculations can be made although the amount of detail required is not as great. Established recoverable reserves at present are approximately 45×10^{12} cubic feet in UK waters, of which 75 per cent are south of 56°N. As the southern area has been largely exploited, it is unlikely that there will be any major additional finds although the rising price of gas could mean that fields previously regarded as uncommercial could be revalued as commercial. The major additions to reserves will come from Scottish waters and, using the same procedure above, estimated ultimate reserves of non-associated gas could be 85×10^{12} cubic feet in UK waters. Also, there are and will be reserves of associated gas in fields such as Brent. Most of the northern oil fields appear to have low gas-to-oil ratios but undoubtedly some of the gas will be commercially exploitable and available in large quantities. It is even conceivable that these reserves of associated gas will be large enough to have a major impact on indigenous energy supplies.

The Maid of Norway's Legacy has turned out to be extremely valuable, for the bulk of UK reserves appears to lie in the East Shetland basin, which makes the UK sector the most prolific in the North Sea. Although not one of the world's largest hydrocarbon areas, on the basis of the estimates of oil reserves above, the North Sea province is an important area in the world context[10] accounting for approximately 2.0 per cent of known recoverable reserves, ranking tenth behind Saudi Arabia (25.2 per cent), Kuwait (12.2 per cent), Iran (11.4 per cent), U.S.A. (6.6 per cent), Iraq (6.0 per cent), Libya (4.9 per cent), Abu Dhabi (4.1 per cent), Nigeria (3.8 per cent) and Venezuela (2.7 per cent).

NOTES TO CHAPTER 3

1. We have used the indefinite article to distinguish this Maid of Norway from *The* Maid of Norway who died in 1290.

2. P. Hinde *Fortune in the North Sea* Henley-on-Thames, Foulis (1966) p. 25.

3. The southern North Sea is defined as that area of the North Sea lying south of the 56°N parallel; the northern North Sea is that area lying north of the 56°N parallel and south of the 62°N parallel. The terms Central Graben and Viking Graben are sometimes used to describe the middle North Sea basin and the East Shetland basin as we have defined them.

4. See pp. 68–75.

5. November 1974.

6. *Production and reserves of oil and gas in the United Kingdom* London, UK Department of Energy (1974).

7. For example, P. R. Odell and K. E. Rosing 'A simulation model of the development of the North Sea Oil Province, 1969–2030' *Energy Policy* (1974).

8. A barrel is the common measurement used in the industry. It contains about 35 gallons and there are between 7 and 8 barrels to the ton, depending on the gravity of the oil.

9. Odell and Rosing op. cit.

10. Excluding the Soviet bloc.

CHAPTER 4

Costs, Opportunities and UK Involvement

There is no high tide without a low tide after it. But, just the same, things came in on the high tide which you could keep when the tide was going out again. (I. C. Smith, *Consider The Lilies*, p. 192)

4.1 INTRODUCTION

In chapter 3 it was shown that the North Sea has proved to be one of the major oil and gas producing areas in the world. Within a few years the energy prospects of the United Kingdom have been transformed, although as yet this new supply of energy is largely untapped. Natural resources are assets that require further development and investment before they can be utilised, and this is particularly true of the North Sea oil and gas reserves. Indeed, the excitement of discovery has tended to obscure the sustained effort and substantial costs that must necessarily be incurred before any tangible benefits can be obtained. The costs must be incurred now and in the immediate future, before the benefits are realised over a longer time horizon.

At present the North Sea is the most costly oil- and gas-producing area in the world, as well as one of the most prolific. Water depths and weather conditions have forced the exploration and production companies to use immensely expensive and often untried equipment. The massive expenditure required and the composition of the exploration groups involved have meant an increasing dependence on external sources of finance, and consequently the high cost of borrowing and the strictures of inflation have added substantially to costs.

These factors are of fundamental importance in appreciating the issues that confront public policy. In chapter 2 we argued that most of

67

the oil fields discovered were heavily 'front-loaded', involving the commitment of substantial capital sums before any revenue is obtained. This fact, which has to be clearly recognised in formulating any taxation regime, is confirmed in this chapter, which investigates in greater detail the expected capital and operating expenditures required in the North Sea. The sums involved are so substantial that they have necessitated a heavy reliance on overseas investment, which again has implications for UK policy in that a taxation regime that does not ensure an adequate rate of return on capital investment will drive away foreign investment and place an immense burden on the UK capital market.

The discussion of financial issues leads on directly to an examination of industrial opportunities opened up and the way in which Scottish and UK industry has responded to the opportunities. There has been a great deal of public comment and interest over the alleged failure of domestic industry to penetrate this new market and the other off-shore activities that are related to it. We shall consider whether this conclusion is or is not justified, the nature of any weaknesses that emerge and the measures that might be adopted to increase UK involvement. Finally, the concluding section of the chapter pulls the various issues together and outlines the major conclusions that emerge.

4.2 EXPLORATION COSTS, CAPITAL INVESTMENT AND OPERATING EXPENDITURES

One outstanding feature of the North Sea developments is the vast size of the outlays required for both exploration and production. In this section we consider the costs of the three phases of activity discussed earlier – exploration, manufacturing and production. A precise definition of the 'size of the market' is difficult because the North Sea is only one of several off-shore areas in which exploration activity is under way. The mobility of various items of equipment – e.g. exploration rigs, supply vessels, lay barges – means that these other off-shore areas present opportunities for exports in the same way that the North Sea can be regarded as an export market for manufacturers in the United States, Japan, etc. It is common for equipment to move from one area to another, and some of the rigs, for example, working in the North Sea have been transferred from the Gulf of Mexico, Nigeria and elsewhere. Also, in the North Sea there are marked seasonal fluctuations in the level of activity and the winter weather conditions drive many rigs and supply vessels away (like sparrows) to more hospitable climates. Similar considerations apply to the downstream end of the industry in that it is difficult to relate refineries and petro-chemical plant to discoveries in the North Sea. Refineries and petro-chemical plant using North Sea oil and gas need not necessarily be established in the United Kingdom and, furthermore, existing refineries could take North Sea oil without increasing capacity by reducing their consumption of oil from other sources. Considerations of this nature will vary so much from company to company and from field to field that it becomes very difficult to assess the overall scale of investment required, or even to distinguish between the 'on-shore' and 'off-shore' markets.

Nevertheless, some generalisations are possible. The obvious one is that developing a North Sea field is much more costly than developing an on-shore field. Although a large proportion of the equipment required is the same, it has to be adapted to the water and weather conditions off-shore and this has proved to be more difficult and expensive than was anticipated originally. Mention was made earlier of the recurring delays in development programmes and one of the major causes has been the need to redesign items of equipment (particularly production platforms) during construction as more information on conditions became available. It would be surprising if such problems did not continue to occur.

Taking the exploration phase first, in Scottish waters it usually takes two to three months and up to £1.5 million to drill an exploration well, although in shallower waters the cost is less. One complicated well has cost the exploration company £5.2 million. Rig hire is by far the major cost item. A few of the exploration companies own their own rigs (e.g. BP's Sea Quest and Shell's Staflo) but in most cases the equipment is on hire or lease from specialist contractors. There is considerable variation in charges for the three types of rig described in chapter 3: a drilling ship normally costs from £6000 to £8000 per day; a jack-up, £10,000–£12,000; and the fleet of semi-submersibles £15,000–£25,000. Of this, some 40–60 per cent represents the capital charge and the remainder the cost of labour, materials, the use of supply boats and helicopters etc. As drilling ships and jack-ups can be used only rarely in the more northerly waters, the necessity of using semi-submersibles largely explains the substantial differences in exploration costs between the various areas. It will be evident also that the loss of even one day's drilling time through bad weather or equipment breakdown can be very costly.

Over the last three years in the northern North Sea the ratio of wells drilled to successful commercial finds has been approximately 8 : 1, which implies an average exploration expenditure of £10 million to discover a commercial field. In the Norwegian sector, exploration drilling began in 1966 and it took four years, thirty-two wells and £50 million to discover the first commercial oil field. For companies that discover commercial fields the costs of exploration and appraisal drilling are obviously small in relation to development costs and anticipated returns, but it should be remembered that there are quite a few exploration groups in the North Sea that have incurred substantial exploration outlays with no return.

Turning to the development programmes, the costs will vary enormously from field to field according to the particular system of production chosen. Unfortunately, of the discoveries listed in chapter 3, only the southern North Sea gas fields are in production and there are only four oil fields in the UK sector – Argyll, Auk, Forties and Piper – for which plans are well advanced and detailed information is available; and hence it is very difficult to draw general conclusions about development costs. In each case, aspects of the development programmes are peculiar to the individual fields. Forties oil, for example, will be piped to a mainland terminal and thence to BP's Firth of Forth terminal; Piper oil is to be piped to an island terminal in Orkney and then transferred into tankers for refining elsewhere; with Auk, the oil will be loaded into tankers at sea from the fixed production platform; and with Argyll the loading will be from a converted drilling rig. In each instance the choice of facilities was dictated by the size of reserves, the expected length of life of the field, the distance from land, the particular geological features etc.

A rough breakdown of the costs of the four fields (at 1974 prices) is given in table 4.1. The main items of expenditure are production platforms and pipelines, which in the case of the Forties and Piper fields account for about 75 per cent of total capital costs. The production platform is the most significant item and the cost will vary according to the number required, size and water depth of the field concerned. On average a steel platform costs between £1000 and £1500 per ton installed on location. Concrete gravity platforms are much heavier but are usually cheaper. For example, the Forties plat-

TABLE 4.1 DEVELOPMENT COSTS OF OIL FIELDS ($£m$)

Forties			*Piper*	
Four production platforms	380		Production platform	65
of which: jackets		(240)	Sea pipeline	65
decks, modules		(60)	On-shore terminal	50
installation		(80)	Development drilling	20
Sea pipeline	90		Other	20
Land pipeline	20			—
Terminal facilities	40			220
Development drilling	65			
Other	55			
	—			
	650			

Auk		*Argyll*	
Production platform	20	Rig conversion	2
Single-buoy mooring	3	Single-buoy mooring	2
Terminal facilities	5	Development drilling	4
Development drilling	8	Other	3
Other	4		—
	—		11
	40		

form which was floated out from Nigg in August 1974 had an installed cost of approximately £95 million (£60 million for the jacket, £15 million for the decks and equipment and £20 million for installation) while the concrete platform for the Beryl field should cost about £60 million installed.

Quite a few fields will need more than one production platform and possibly separate platforms for pumping, separation and accommodation (as with the Frigg gas field), and obviously until sub-sea completions establish themselves this will remain the dominant sector of the market. The requirements for modules, generating equipment etc. are closely linked to platform requirements.

Pipelines similarly account for a large proportion of capital costs. The cost of the pipe itself is overshadowed by the costs of laying pipelines off-shore, which are in the range of £400,000–£800,000 per mile depending on water depth and seabed conditions. As can be seen from table 4.1 the 110-mile pipeline from the Forties field to the mainland is expected to cost £90 million and the 135-mile pipeline from the Piper field to Orkney will cost around £65 million. Pipe-laying on-shore is also expensive, at about £100,000 per mile. In comparison, the costs of the pipeline itself – about £50,000 per mile – and the provision of protective coating – about £35,000 per mile – are relatively small.

In fact, it is in this area of platform installation and pipe-laying that the pressure on both costs and technological advance has been most severe. A completely new generation of derrick barges, pipe-laying and pipe-burying barges has had to be designed and constructed for North Sea conditions and this has been reflected in the supply and cost of the suitable equipment. The latest pipe-laying barge, for example, has cost £40 million – twice as much as the new generation of semi-submersible rigs – and hence its daily hire charge is enormous.

Consequently for the smaller fields and those with short life periods – such as Argyll and Auk – the use of off-shore loading terminals and tankers is a more attractive proposition. Expected rates of return are not sufficient to justify substantial capital outlays, albeit at the expense of much greater operating costs. The Argyll field is in fact using a converted exploration rig instead of a platform, which, together with a single-buoy mooring system, means a capital cost of around £11 million. As with exploration rigs, the oil and gas companies almost invariably prefer to hire tankers, supply boats and other transport facilities (on occasion from sister companies in their own operating groups) to reduce capital expenditure.

On-shore, it is very difficult to detail costs because the facilities required vary considerably. All fields require a landfall terminal and refining capacity but in many cases these exist already and may or may not need expansion. The terminal facilities being constructed specifically for the Forties and Piper fields will cost £49 million and £50 million respectively, but in both these cases there is provision for future discoveries to be fed into the system. In Shetland work is proceeding on a large multi-user terminal and pipeline complex to accommodate many of the fields in the East Shetland basin and this is likely to be an increasingly common practice. Finally, there is the question of refinery costs. On the oil side, a majority share of North Sea oil is likely to be refined outside the United Kingdom, and within the UK there are no signs as yet of new refining capacity being constructed to take North Sea oil, although there are refinery proposals outstanding in Scotland which range from £50 million to £200 million.[1] On the gas side, processing plant is much more easily identified with specific North Sea discoveries but the only development in the foreseeable future will be the Frigg gas terminal at St Fergus which is being constructed at an estimated capital cost of £85 million, in addition to the distribution and pipeline costs to the national network.

All the above should be regarded simply as giving a rough indication of the magnitude of the capital costs involved. These costs will vary enormously from field to field, as will operating costs. Normally, annual operating costs are between 3 and 5 per cent of the total capital expenditure (i.e. about £30 million per year for Forties), but the Piper costs could be up to £20 million per year (about 8 per cent) and Argyll will be around £15 million per year – which is greater than the initial capital outlay. Hence it is even more difficult to generalise about operating costs which in total will probably reach £100 million per year from 1976 onwards.

Nevertheless, it is necessary to provide some indication of the likely level of capital expenditure required for the UK sector of the North Sea and the simplest method is to calculate the average capital cost per unit of offtake and to apply this average to the forecasts of production rates given in chapter 3. Unfortunately, reliable information is available only for the four fields in table 4.1 and the Argyll and Auk fields have very low capital costs. For Forties a peak production rate of 400,000 barrels per day implies capital expenditure of £1625 for each barrel per day of peak production; for Piper a peak production rate of 225,000 barrels per day implies expenditure of £980 for each barrel. Larger fields such as Ninian and Brent, however, are likely to be more costly and it seems that on average development costs will be between £1400 and £1500 (at mid-1974 prices) for each barrel per day of peak production; i.e., a field producing a peak flow of 100,000 barrels per day would cost between £140 million and £150 million to develop. It should be stressed that this is just an average figure.

Using this range, and applying it to the forecast offtake rate from the United Kingdom sector of three million barrels per day by 1980, implies capital expenditure of between £5000 million and £6000 million by that year, with an ultimate expenditure requirement of around £15,000 million, the time-scale of which would depend on future production plans. To this must be added development expenditure for the gas fields, which is likely to total £1500 million by 1980, and expenditure on exploration drilling, which is estimated at around £1800 million for the same period (on the basis of the forecasts in chapter 3 of the number of wells to be drilled).

TABLE 4.2 ESTIMATES OF EXPENDITURE : UK SECTOR*
(£m)

	up to end 1973	1974–80	Total to 1980
Oil developments	500	5300	5800
Gas developments	100	1400	1500
Exploration	300	1500	1800
	900	8200	9100

*Excluding operating costs

These estimates are summarised in table 4.2, which covers exploration costs and capital expenditure (but not operating costs). The figures represent the mid-points of the various estimates and should be regarded only as giving a rough indication of the magnitudes involved. Of the £900 million incurred to the end of 1973 about 65 per cent took place in or for the northern North Sea and this proportion should rise to 95 per cent by 1980, reflecting the massive buildup of activity in more northerly waters.

By detailing the time-scales of individual field programmes it is possible to give some indication of the yearly breakdown and this has been attempted in table 4.3. Again, the figures should be regarded only as giving a rough breakdown of the cost over time as the lumpy nature of programmes necessitates a rather arbitrary allocation for particular years. It will be seen from the table that we expect an annual average rate of expenditure of around £1200 million in the UK sector for the rest of the present decade, with the years 1975–77 embracing the peak of development expenditure. Throughout the 1980s annual expenditure should average around £700 million, although the crucial (and unclear) variables here are government control over production and exploration and future demand for hydrocarbons.

This scale of expenditure is massive. It may be difficult to put such vast sums in perspective but it represents a potential 25 per cent increase in private capital expenditure in the United Kingdom. In 1973 private industry's capital expenditure in the UK was just over £4000 million. Furthermore, it must not be forgotten that there are other sectors in the North Sea that are potential export markets for the domestic off-shore industry – as indeed are most of the off-shore areas of the world, particularly for mobile equipment such as rigs and supply vessels. A government-commissioned study which was

TABLE 4.3 EXPENDITURE ESTIMATES, 1973–80*

(£m)

	Exploration	Oil development	Gas development	Total
1973	100	300	50	450
1974	150	600	150	900
1975	225	800	225	1250
1976	300	900	250	1450
1977	275	850	250	1375
1978	225	800	200	1225
1979	200	700	200	1100
1980	125	650	125	900

*Excluding operating costs

published in 1972[2] estimated that the UK sector represented 25–30 per cent of the world off-shore market, with other North Sea sectors adding another 10–15 per cent. Although these proportions may seem high in the context of the high level of off-shore exploration throughout the world, the main point to bear in mind is that the North Sea is the most costly area to date which the oil and gas industries have had to develop. This is exemplified by the capital costs involved: as was estimated above, capital costs for North Sea oil fields are in the range £1400 to £1500 for each barrel per day of peak capacity; the equivalent figure for the Gulf of Mexico is approximately £900; for Nigerian water £400; and for the Middle East £40. As discoveries are made in the more northerly and deeper waters of the North Sea these cost differentials will widen.

Finally, as with the estimates of reserves and production rates in chapter 3, we should point out that our estimates of costs and the size of the market are higher than those set out in the study mentioned above, undertaken by the International Management and Engineering Group (IMEG), which forecast that the average annual rate of expenditure in the UK sector during the present decade would be around £300 million, compared with our estimate of up to £900 million. Part of the difference can be explained by inflation (the IMEG figures being at 1972 prices), but it is largely a matter of differing views of the future. Our own estimates, however, are very much in line with those produced by various banks and stockbrokers[3] and, indeed, IMEG's estimates for 1973 and the first half of 1974 represent only 40 per cent of the actual expenditures which were incurred.

4.3   SOURCES OF FINANCE

The discussion above has ignored the problems of financing the massive scale of exploration and development that is envisaged for the North Sea. Capital, like labour and equipment, is not in unlimited supply and it is necessary therefore to look more closely at potential sources of finance.

A crucial factor is the nationality of the companies involved because this in turn is one of the main determinants of the origins of capital. As is well known, the oil and gas industries are dominated by the US majors and they are present in large force in the North Sea. Of the oil discoveries listed in table 3.3 (p. 61), the United Kingdom share – both public and private – is just over 44 per cent of total recoverable reserves, with the US share slightly lower (38 per cent) and most of the remainder in the hands of French, Norwegian and Canadian companies.

When it comes to the provision of finance US dominance is likely to be even greater. The sums mentioned above are far in excess of what the domestic capital market can provide and at the present time Eurodollar and European Community funds are unable to provide a significant proportion. Banks in the United States, on the other hand, have considerable experience of the oil and gas industries and a long history of dealing with the major companies; and in addition they have access to funds of the required magnitude. Consequently, American banks and financial institutions have played the major role to date in providing finance for development programmes. This will undoubtedly continue although it will be interesting to see to what extent surplus oil revenues from the Middle East will find their way to the North Sea. As regards US finance, BP have raised a loan of nearly £400 million for Forties, approximately 60 per cent coming from US sources; the Ekofisk consortium have borrowed £80 million from the US Export-Import Bank – this largely tied to the purchase of American equipment; Placid International Oil have raised 80 per cent of a £70 million loan for gas developments in the Netherland sector from US sources; and even the UK drilling company Kingsnorth Marine Drilling have raised 80 per cent of an £18 million loan in the US for the purchase of two rigs.

Although we are still at the early stages of development, three patterns are emerging in addition to that of US dominance. Firstly, there is the need for external financing. In the past a great deal of exploration and development was financed from the internally generated funds of the oil companies themselves. In the case of small fields such as Auk and Argyll this is still possible but generally the size of the funds required and the multiplicity of projects (Shell/Esso, for example, in addition to the Auk and Cormorant fields, have shares in the Brent, Dunlin and Forties fields) are forcing the companies to turn much more to external sources. Secondly, there is the increasingly popular system of production payment financing under which loans are amortised by the oil produced. This is the principle of the Forties loan, which sets up a special company to buy the oil from BP at the wellhead and then resell it at a price sufficient to repay the loan. Thirdly, there is the tying of loan finance to the purchase (or lease) of equipment – as in the case of Ekofisk and a £42 million loan to the Total group for the Frigg gas field from a group of UK banks. This will tend inevitably to reinforce US domination and the consequences of this are discussed in more detail below.

Lastly, given the present economic climate throughout the Western world it is essential that the North Sea continues to remain an attractive proposition to both oil companies and financial institutions. Recent uncertainty about government policy has forced certain bodies to hold plans in abeyance until a clearer indication of expected profitability and rates of return emerged. Doubts about the future price of crude oil reduce the collateral security of North Sea production and there is increasing interest in other off-shore areas, particularly those nearer the American market.

In the UK finance has come mainly from the clearing and merchant banks – the three Scottish clearing banks put up £200 million for a variety of off-shore and on-shore projects – and investment and unit trusts. One very important aspect is the large number of smaller companies involved with the exploration groups and consequently with some of the major discoveries. Part of the Ninian field, for example, extends into block 3/8 which is licensed to a group consisting of BP (50 per cent), Ranger (20 per cent), London and Scottish Marine Oil (15 per cent), Scottish Canadian Oil and Transportation (7 per cent), Cawoods Holdings (3.75 per cent) and National Carbonising (3.75 per cent) – and there are many similar groups. The smaller companies in some of the consortia will have to raise finance far in excess of their market capitalisations and this raises tricky problems. New share issues are unlikely to be successful at the present time, as a few companies have discovered to their cost.

In conclusion, our own feeling is that the United Kingdom will only be able to provide between 30 and 35 per cent of the total finance required (a slightly higher share than at present) and that most of the remainder will come from US sources – and possibly from surplus Middle East revenues. Given the factors outlined above, there appears to be little that could be done to increase the domestic share and in view of our continuing balance of payments problems an inflow of foreign capital is very welcome, as is shown in chapter 5.

4.4 SCOTTISH AND UK INVOLVEMENT

We have tried to demonstrate that the size of discoveries in the North Sea and the financial and industrial opportunities are massive. From the domestic point of view, however, the situation can be likened to that of the man who finds he has twenty-four points on the football pools, expects a dividend of £500,000, and then discovers that he has to share the pool with five hundred other lucky winners. For various reasons, both avoidable and unavoidable, the Scottish and UK share in the North Sea developments will be much less than the total sums estimated above. It cannot be denied that our involvement to date has been disappointing and this has been one of the major political issues in the last few years.

In 1971 the Department of Trade and Industry (as it then was) commissioned the International Management and Engineering Group (IMEG) to undertake a study to assess the potential scope for participation by UK industry in the supply of equipment, materials and services required by the off-shore oil and gas industries. In their report, published in 1972, IMEG described the situation in the following terms:

> British industry has to break into an area of activity in which foreign enterprise, mainly American, has established an entrenched position in off-shore work, and is daily strengthening that position by the accumulation of experience in solving the still more difficult problems of the North Sea. . . . The implications of off-shore engineering and contracting are much wider than the scale of the work itself, running through the whole of British industrial effort. The sources of supply of equipment and materials of all kinds to

go on or in the major facilities rests essentially on who specifies and who buys. At present, these functions are mainly in the hands of United States and United States-experienced engineers who tend naturally to favour proven equipment from known and trusted suppliers. This, therefore, is the situation which has to be changed. . . .[4] [The report stressed the] unwillingness of the oil companies to risk inadequate performance by a contractor engaging in work not previously undertaken. This problem is magnified by the size of the investments and by the extreme effect on the oil companies' profits which can be caused by a comparatively slight delay in completion.[5]

IMEG estimated that at that time the share of UK enterprise in the United Kingdom off-shore market was about 25–30 per cent of the total and that unaided the share would probably increase to 35–40 per cent of the total by the late 1970s. They put forward certain recommendations, however, which they felt would increase the share to some 70 per cent.

Before considering IMEG's recommendations and the response to them, it is essential to look at Scottish and UK involvement in some detail. Overall, the situation is dispiriting although in some areas the performance has been encouraging. The crucial areas – such as contracting – remain firmly in the hands of foreign enterprise and these are the areas where the problems are virtually intractable. A detailed investigation of all the sectors involved would occupy a disproportionate amount of this book, so attention will be concentrated on four key sectors: project management, the contracting and construction of exploration rigs, the construction of production platforms and pipeline construction and laying.

The most crucial is probably project management in that most of the exploration companies have little expertise in project engineering and construction and usually contract this out. If an overall project manager is appointed then this company will have considerable influence over the ordering of production platforms, pipelines, generating equipment etc. The field is heavily dominated by United States concerns which is somewhat surprising because quite a few United Kingdom companies have had considerable experience both in the domestic market (e.g. with refinery construction in the late 1950s) and overseas, particularly in the Middle East. In the last few months three UK companies have been appointed project managers for massive oil developments in the Middle East but none of these companies is much involved in the North Sea. There are some encouraging signs – e.g.

the involvement of CJB Offshore and Taywood Seltrust in design and engineering work for the Thistle field – but this aspect is so important that a great deal more has to be achieved.

Regarding the contracting of exploration rigs, at the end of 1973 there were 246 rigs operating off-shore throughout the world. Of these, 197 were United States-owned, the major owners being ODECO (23), Offshore Company (19), Penrod (14), Global Marine (13), Sedco (13) and Zapata (13). Of the remaining 49, 6 belonged to joint United States–European ventures and only one was United Kingdom-owned (BP's Sea Quest rig), although the Royal Dutch–Shell group (which has a substantial British interest) owns 5 rigs.[6] In the North Sea in mid-August 1974, 29 rigs were operating in the United Kingdom sector: 19 United States-owned, 4 United States-European, 3 Norwegian, 1 French, 1 Royal Dutch–Shell, and BP's Sea Quest.

The position is similar regarding rigs under construction, with over 70 per cent of the rigs being built destined for United States or joint United States–European owners. There are signs, however, of a few United Kingdom companies making serious attempts to enter the market. Christian Salvesen have purchased a converted drill-ship which is presently operating off north-east Australia; four companies – Kingsnorth Marine (2), Atlantic Drilling (2), Celtic Drilling and Wimpey – have semi-submersibles on order and three other companies – Ben Line, Holder Line and Blandford Shipping – have entered joint ventures with United States or Norwegian interests. Kingsnorth Marine, Atlantic Drilling and Celtic Drilling all have substantial Norwegian participation and were established in fact at the instigation of the Norwegian Stott–Nielsen group. Of the others, the most interesting venture is probably the establishment of Ben–ODECO, with equal shares being put up by ODECO, the largest United States exploration rig concern, and Ben Line Offshore Contractors, in which Ben Line have the majority shareholding and North Sea Assets and the Royal Bank of Scotland have smaller interests. Ben–ODECO have already taken over the ODECO jack-up Ocean Tide.

Ocean Tide is one of the few exploration rigs that have been built in the United Kingdom. At the moment only two shipbuilding yards – Marathon Manufacturing and Scott Lithgow, both on the Clyde – are building rigs (of the jack-up and drill-ship types). Marathon are wholly United States-owned and took over the old John Brown yard in 1972. Prior to that, only twelve rigs had ever been built in United

Kingdom yards, dating back in fact to 1958. Nine of these were jack-ups but three semi-submersibles were built – Ocean Prince (Teesside, 1966), Staflo (Furness, 1967) and Sea Quest (Belfast, 1966). Ocean Prince, incidentally, sank in 1970, a year after the Constellation, a jack-up which was built by John Brown on the Clyde in 1966. Indeed, the Clyde has an ignominious history of rig-building because when the last rig to be built there (prior to Marathon) was being towed down the river, the legs dropped off!

IMEG considered in some detail the role of the United Kingdom shipbuilding industry with regard to rig construction and concluded that there were no major drawbacks to the construction of jack-ups and drill-ships but that semi-submersibles required significant improvements in operation, the most important of which was the acquisition of suitable craneage. Their report made various suggestions as to how these improvements could be achieved,[7] but little account has been taken of them.

This is in marked contrast to what has happened in Norway in recent years. Despite the relatively low level of activity in the Norwegian sector, Norwegian involvement in rig contracting and construction has built up steadily and is now four to five times greater than that in the United Kingdom. The outstanding example is that of the Aker group, which is currently building twelve semi-submersible rigs in Norway with another twelve of their design being constructed on licence in yards in Finland and Japan. The Aker group moved into the business only in 1965. Given that Norway and the United Kingdom, particularly Scotland, have similar maritime traditions, this sector represents the prime indictment of UK enterprise in the North Sea developments.

The construction of production platforms gives rise to similar criticisms. It was suggested earlier that by 1980 the market for platforms for the United Kingdom sector will total approximately £1300 million. At the time of writing, sixteen platforms have been ordered for oil fields in the UK sector. Six of these platforms – costing £125 million – are being built abroad: three in Norway, two in France and one in the Netherlands. With four large steel platform builders located in Scotland and northern England, there appeared to be little chance of steel orders going abroad, but inability to meet delivery dates has resulted in the latest order – for the Claymore field – being placed in France. Furthermore, the switch to concrete platforms has found the United Kingdom largely unprepared. Only one concrete site is in operation here – the McAlpine site in Argyll – and consequently five of the eight orders that have been placed for concrete platforms for the UK sector (i.e. including the Frigg gas field) have gone abroad, four to Norway and one to the Netherlands. Four of the Norwegian orders have been for platforms of the Condeep design, which is the type of platform the Mowlem–Taylor Woodrow consortium wanted to build at Drumbuie in western Ross, planning permission for which site has been refused by the Secretary of State for Scotland after a lengthy public inquiry. The implications of the Drumbuie decision are considered in chapter 7 but some of these were made clear by the Department of Energy in their announcement that platform sites in Scotland would be taken into public ownership. In announcing this policy the Secretary of State for Energy stated that:

> . . . . we should build as many platforms as we can in this country. The government attaches great importance to encouraging United Kingdom industry to become an international competitor in the off-shore market. Platforms are the key items of this market and the most expensive . . . . In general terms, the subcontracts for specialised equipment for the structure and the deck tend to follow the award of the main contract. So a British platform order means up to £60 million worth of business to British industry.[8]

An encouraging feature of platform construction, however, has been the involvement of UK companies in joint ventures with United States, French, Italian and Norwegian companies with extensive experience in off-shore engineering. Four of the five groups building platforms are joint ventures: Highlands Fabricators, comprising the United States giant Brown and Root and the United Kingdom construction company George Wimpey; RDL North Sea, comprising the United Kingdom steel fabricators Redpath Dorman Long and an Italian group made up of Saipem, Micoperi, Fiat, Pirelli etc; Laing Offshore, a joint venture between another UK construction company, John Laing Construction, and the French group ETPM; and McAlpine, comprising Sir Robert McAlpine and Sons and the French designers Sea Tank. Joint ventures of this type are an obvious way of entering a market dominated by foreign companies and should set an example for other sectors of the off-shore oil and gas markets. This was one of the main recommendations of the IMEG report, which added that 'partnerships with foreign firms whether in contracting or in equipment supply should be framed to enable the British partner to acquire full knowhow and to be free to establish itself in world export markets without limitation by the foreign partner's independent interests'.[9]

Another sector in which partnerships are vital is pipe-laying. The potential market here is enormous – around £900 million up to 1980 – but the sector is dominated by five foreign companies. United Kingdom companies have been reticent in investing the £15 million to £25 million which a pipe-laying barge costs and to date the only UK interest is a 25 per cent share held by North Sea Assets and the Bank of Scotland in a barge which should be in operation in 1975.

Similarly, most of the steel required for the off-shore pipelines has been ordered abroad. All the off-shore pipeline for the Forties, Piper and Ninian oil fields is coming from Japan, as is that for the Frigg gas field, and the pipeline from the Norwegian Ekofisk field to Teesside is being manufactured in West Germany and the United States. At present the British Steel Corporation does not have the plant to manufacture pipe of the required thickness for the underwater pipelines and it seems that it will be 1976 or 1977 before such plant comes into operation. British Steel decided some years ago to concentrate on the land pipelines and the steel casing needed for wells and has a good record in these fields. Undoubtedly, however, it underestimated the future North Sea demand for off-shore pipelines.

All these are examples of major sectors of the North Sea developments in which UK industry has missed opportunities. Other examples can be instanced easily enough – e.g. supply boats, development drilling, well logging – although, to be fair, in certain sectors Scottish and United Kingdom involvement has been admirable. British Airways Helicopters and Bristow have quickly and efficiently dominated the helicopter side of off-shore transport; companies like Seaforth Maritime in Aberdeen are diversifying into a wide range of supply activities; module construction is dominated by four UK enterprises; and a similar situation exists in the manufacture of pumps, generators, wire rope etc.

Overall, however, an analysis of purchasing patterns to date indicates that approximately £525 million of the estimated £1600 million development expenditure to end 1974 will have gone to Scottish and UK companies. This share represents 32 per cent of estimated total expenditure, a proportion that is only marginally higher than that estimated by IMEG in 1972. In its own right this performance is disappointing, but it is even more so when seen in the light of the vast export markets appearing elsewhere as the off-shore search for hydrocarbons spreads throughout the world.

4.5 CONCLUSIONS

In chapter 3 we set out the vast scale of the oil and gas discoveries in the North Sea. In this chapter we have attempted to show the scale of effort required to bring the discoveries into production. This is undoubtedly massive – close to £10,000 million by 1980 – and creates considerable problems in meeting timetables and obtaining finance.

The North Sea is the most difficult and costly off-shore area that the oil companies have had to develop to date. The fourfold rise in the price of oil has made the area much more attractive and it would be shortsighted to disregard the implications of changing oil prices and related external considerations. Much of the effort being under-taken is viable only at the existing price level and it must be recognised that a substantial fall in oil prices would have far-reaching effects on the scale of activity in the North Sea. Development and operating costs in the North Sea are much higher than in the Middle East – in some cases by a factor of 30 – and although the margin between production costs and selling prices is substantial the North Sea is inevitably a high-cost area. To a small extent North Sea oil is protected by virtue of its high-quality, low-sulphur content and proximity to west European markets but this protection probably only amounts to around $1 per barrel. It is essential therefore that a stable and attractive environment is maintained in the North Sea over the next decade.

As regards the involvement of Scottish and UK industry, the crucial questions are why this involvement has been so poor and what can be done to improve the situation. Undoubtedly, many of the problems stem from the country's disappointing economic record over the last few years and from the low level of investment in the manufacturing sector. The overriding problem has been the inability or unwillingness of industry to take the risks necessary to become involved and it is in this sphere that comparisons with Norwegian experience are most appropriate.

In this context there is some dispute about the importance of the rate of exploration and development. On the one hand it can be argued that a much slower rate of development in the UK sector would give domestic industry time to equip itself for a considerably larger and more consistent involvement. On the other hand the argument is that it is only in a large and fast-growing market that British industry will be able and willing to break the virtual monopoly of foreign competitors. The former argument – that of the infant industry – is a strong and attractive one. It appears to have been adopted with some success in Norway, although there are many differences between the Norwegian and UK situations. Certainly in most spheres of activity the pressure on the oil companies to obtain equipment and services quickly has inevitably meant that they have had to turn to suppliers with existing facilities and experience – in other words to overseas, mainly US, companies involved in the Gulf of Mexico etc. Given a longer learning time, it is possible that UK companies could have equipped themselves with sites, plant and skills to compete with such companies.

It is impossible to reach a firm conclusion on this matter but our own view at the present time is that the arguments against an infant industry policy are stronger. In the first case it is rather misleading to describe certain parts of the off-shore industry as infant. A good deal of the equipment is common to traditional processing and maritime industries and has been a familiar output of UK industry in the last decade. As was shown in chapter 3, exploration activity has been under way in the North Sea since 1964 and the peak of the southern North Sea gas development phase was in the late 1960s. Over the last decade there have been two periods of two to three years during which the level of activity and interest was such that UK companies had outstanding opportunities to establish strong competitive positions, and these opportunities have been largely missed. Secondly, it must be remembered that the North Sea is only one of a growing number of off-shore hydrocarbon areas and that it is very desirable that we should establish a strong presence in other areas. There is a real danger that if the UK were to adopt a protectionist policy towards its own sector then other countries would react similarly. This could also result in a much greater reduction in foreign interest in the North Sea than might be anticipated. Thirdly, an infant industry policy is essentially incompatible with the government's desire for quick exploitation over the next few years.

Our own view is therefore that, bearing in mind the desirability of achieving an important UK presence in other off-shore areas, UK involvement in the North Sea would not be significantly improved by slowing down the rate of development and that the main problems to date have not been related to the level and time-scale of activity but rather to risk-taking and performance. It is in these latter areas that there is considerable scope for joint industry and government action.

In particular, if government policy urges a rapid rate of exploration and exploitation, then a prerequisite is that on-shore conditions are suitable for UK industry to participate. The best example is that of platform construction, orders for which have gone to Norway and other countries because of a shortage of sites with planning permission in Scotland. As platform orders are crucial to development programmes there has been a serious loss in income and employment to the UK and this will continue until a framework is created which allows domestic companies to compete on equal terms with those overseas.[10] Similarly, a constraint on development in north-east Scotland and the Highlands and Islands has been a lack of public expenditure on infrastructure, particularly housing and communications.[11]

On the financial side there appear to be two factors that have to be improved: the availability of cheap credit and the provision of venture capital on a massive scale. Regarding the former, British companies have had to compete with foreign suppliers on very unfavourable terms but the government has introduced an interest relief scheme which provides grants at 3 per cent per year on credit to finance contracts for providing UK goods and services. This scheme was designed to counteract preferential interest rates available to foreign suppliers and to underlay the Frigg loan mentioned above. Progress on the latter factor (the provision of venture capital), however, has been non-existent. In most aspects of contracting the initial investment required (e.g. the purchase of heavy cranes or pipe-laying barges) is far beyond the capabilities of individual companies, but entry into the market would be possible for a group of companies or for a larger company if the risk was insured by government action. What is needed is a body on the lines of the old Industrial Reorganisation Corporation to promote mergers or joint ventures and to provide capital in areas where it is needed. Such thinking was behind the major IMEG recommendation of the urgent establishment of a Petroleum Supply Industries Board. The government response was to set up the Offshore Supplies Office, with the stated objective of increasing the participation of UK industry, but this body has played a very low-key role since its inception and operates along the lines of a local authority development agency rather than the Industrial Reorganisation Corporation.

IMEG believed that if their proposals were accepted the UK share in its own off-shore market would rise to about 70 per cent; under closer examination this seems an unrealistic target but, given sensible action to solve the problems described above, there is no reason why the share could not be increased to between 45 and 50 per cent from its present level of 32 per cent. If this could be done in such a way as to create a strong and competitive off-shore industry in the UK, when the North Sea tide goes out we could turn our attention to other off-shore areas.

NOTES TO CHAPTER 4

1. The refinery situation is discussed in more detail on pp. 115–6.
2. *Study of potential benefits to British industry from offshore oil and gas development* London, Department of Trade and Industry (1972).
3. See for example Wood, Mackenzie and Co., *North Sea Service* Edinburgh (1974) and *The U.K. Continental Shelf Oil Resources* London, London and Dominion Trust (1974).
4. *Study of potential benefits* op. cit. p. 7.
5. ibid p. 29.
6. *Petroleum Economist* (August 1974) p. 304.
7. *Study of potential benefits* op. cit. pp. 85–9.
8. *The Scotsman* (13 August 1974).
9. *Study of potential benefits* op. cit. p. 6.
10. This issue is discussed in more detail in chapter 7.
11. See chapter 7.

CHAPTER 5

The Economic Significance of North Sea Oil

Wine that maketh glad the heart of man: and oil to make him a cheerful countenance. (Psalm 104, v. 15)

5.1 INTRODUCTION

This chapter contains a good deal of calculation and computation which many may find tedious. We must ask for the readers' forbearance. The arithmetic may be dull, but the conclusions to which the arithmetic points are of very considerable economic significance. We shall assess the nature and the extent of the economic benefits that are likely to arise from the North Sea oil and gas discoveries and their exploitation, described in the two preceding chapters. Such an assessment is crucial if we are to understand the nature of the industry and the form in which the major benefits from oil will accrue. This, in turn, will set the limits and possibilities of government action and economic policy.

The task of assessment is not an easy one. Rational calculation is more difficult and less exciting than guesswork, but it is a more reliable guide to evaluation and policy decision. So we have attempted to set out the basis of our assessment clearly. Wherever possible we have tried to quantify our forecasts and state the assumptions and evidence on which they are based. On occasion it may be felt that we have made too liberal a use of the 'back of an envelope' beloved in economists' folklore. Our defence would be that this is an advance on the back of a postage stamp, which seems to have formed the canvas of much of the speculation to the present date!

We believe that our assumptions are reasonable and realistic, given the basis of current knowledge. Inevitably, there will be room for disagreement. Hence, the argument is set out in a fashion that will allow the reader, if he so desires, to substitute alternative assumptions and consider how these might modify our conclusions. We do not believe that these modifications will prove to be substantial provided two central assumptions hold. These are that the price of crude oil remains at mid-1974 levels in *real* terms, and that our estimates of the production rate of North Sea oil and gas are realised. Even if we relax these assumptions somewhat, this would modify rather than overturn our main conclusions. 'Slippage' in production targets would delay, but not eliminate, the achievement of the economic benefits that we have attempted to measure. Again, while a reduction in the price of oil would certainly reduce the value of those benefits, it seems extremely unlikely that this would overturn our conclusion that the chief economic benefits from oil will accrue through the current account of the balance of payments and to the Exchequer in the form of increased tax revenue. Compared to these effects other economic benefits, for example the direct impact on employment and incomes, are of only minor economic significance.

In the remainder of this chapter we shall examine likely developments in the supply and demand for energy and from this estimate domestic energy requirements to 1980. We can then obtain a view of the effect of North Sea oil and gas production on the current account of the balance of payments. To complete the balance of payments calculations we have to consider certain further issues – the flow of capital to finance North Sea oil activities; the purchase of equipment and services from overseas; the import content of operating expenditures; the government 'take' in the form of royalties and taxation and the share between profits and interest remitted overseas and profits and interest accruing to British residents. This also provides a view of likely government revenue from oil and a means of establishing the overall effects on the balance of payments. Finally, we summarise the main conclusions.

5.2 Energy: Supply Potential and Future Requirements

UK North Sea oil production potential was calculated in chapter 3. In brief, we consider that oil production may reach 3170 million barrels per day by 1980 and, even on conservative assumptions, we would expect production potential to remain at that level throughout the 1980s. Natural gas production from the southern North Sea appears likely to reach a plateau of some 4000 million cubic feet per day (m.c.f.d.) in 1975.[1] Official estimates[2] suggest a flow of 5000 m.c.f.d. for the total UK sector in the late 1970s.[3] Although these official estimates may prove to be on the low side, the degree of understatement will be less than that for oil as most northern oil fields appear to have a low gas-to-oil ratio. For the purposes of illustration, therefore, we have assumed that the flow of natural gas will increase from some 3000 m.c.f.d. in 1973, by 500 m.c.f.d. annually to reach a plateau of 5500 m.c.f.d. in 1978. The detailed estimates are shown in table 5.1. To facilitate comparison of the respective values of the potential production rates the natural gas estimates have been converted to million tons of oil equivalent (m.t.o.e.). This shows that by 1976 the flow of North Sea oil and gas will make a similar contribution to our energy requirements, but by 1980 the estimated production of oil is four times the oil equivalent of the estimated flow rate of natural gas.

TABLE 5.1 ESTIMATED OIL AND GAS PRODUCTION, UK NORTH SEA

	Oil		Natural gas	
	'000 barrels/ day	million tons oil per year	m.c.f.d.	m.t.o.e.
1974			3500	31
1975	120	6	4000	35
1976	565	28	4500	39
1977	1200	60	5000	44
1978	1825	91	5500	48
1979	2760	137	5500	48
1980	3170	158	5500	48

Conversion factors: barrels/day × 49.8 = tons/year
10,000 m.c.f.d. = 0.24 million tons oil equivalent (m.t.o.e.)

In 1973 total UK energy consumption of primary fuels amounted to 202 million tons of oil equivalent (m.t.o.e.) with the market shared as follows: oil, 47 per cent; coal, 39 per cent; natural gas, 11 per cent and nuclear and hydroelectricity, 3 per cent. The future growth of demand depends crucially on two factors – the growth rate of GDP and the 'energy coefficient' (the ratio of growth in total energy requirements to the growth of GDP), which in turn depends on likely savings and improved efficiency in energy consumption. We have estimated the growth of total UK primary energy consumption by assuming:

1. a GDP growth rate of $2\frac{1}{2}$ per cent per annum which is in line with the growth rate of GDP over 1970–74 and appears likely to be an over-optimistic rather than a pessimistic prediction of the realised outcome over 1974–80;

2. an energy coefficient of 0.7 over 1973–75, falling to 0.6 over 1976–80.

The estimated increase in total energy requirements resulting from these two assumptions is shown in the final column of table 5.2 below. It is worth noting that the estimated increase in requirements over 1973–80 is 11 per cent compared with an increase of 15 per cent in the six-year period to 1973.

Our next task is to calculate the share of the total market which will be taken by the various fuels. Whatever the realised outcome, it will be very different from any forecast outcome made in the late 1960s or early 1970s. The increase in crude oil prices, the discovery of additional indigenous supplies of natural gas and undreamt-of reserves of oil, and the comparative failure of the nuclear electricity programme provide entirely new perspectives against which we have to reconsider future developments. These will depend on demand and supply possibilities:

> The future demand for coal depends on price . . . ; price turns on costs of production and delivery; costs on how much coal has to be produced . . . which in turn depends on the demand. The circularity is complete.[4]

The interdependence of the different producers further complicates the problem. Demand and supply projections for one market depend on the assumptions made for other sectors. For example, the future demand for coal depends on what estimates are made of the likely availability and the price of oil, and vice-versa. Somewhere one has to break into this system of interlocking circles, and making vigorous use of Occam's Razor we have constructed our detailed estimates of future energy requirements on the following basis:

1. We assume no significant increase in hydroelectricity generation.

2. Nuclear power projections up to 1977 can be based reasonably firmly on the capacity of existing stations, or those currently under construction. The difficulties encountered with that programme,

reflected in the 1974 announcement of a limited construction programme for steam-generating, heavy water reactors, indicate only
a modest further increase in capacity to 1980. This is reflected in
table 5.2, which suggests a total output of nuclear and hydroelectricity of only 13 million tons of oil equivalent by 1980.
3. With natural gas the balance of considerations would seem to
point, as in the 1960s,[5] to quick depletion, and this implies the
maintenance of a pricing policy which will continue to give natural
gas a competitive edge over other fuels. We would expect, therefore, that natural gas will further penetrate both industrial and
domestic markets and that the whole of the potential supply shown
in table 5.1 for the UK sector of the North Sea will be taken up,
together with imports of natural gas from the Norwegian sector
of Frigg, which is taken at 1000 m.c.f.d. in 1978–80.

We believe that the above forecasts will prove to be reasonably
accurate, but it is more difficult to predict how the remainder of the
primary energy market will be shared between oil and coal. The
recent rise in oil prices would be expected, other things being equal,
to make coal much more competitive. Unfortunately the *ceteris
paribus* condition does not hold. Coal is easily the most labour-
intensive of the energy industries and the course of wage settlements
since the 1972 Wilberforce awards, together with the disappointing
productivity record, makes the future price of coal as difficult to
predict as the future price of crude oil. We have assumed, for want
of anything better, that coal will continue to have some price advantage
over oil, but that it will prove difficult to maintain the output of coal
at present levels. Hence,

4.   It is expected that the decline in coal output will continue, but
more slowly than in the past, so that by 1980 output will be 120
million tons of coal per annum, or 71 million tons of oil equivalent.[6]
5.   Oil is therefore treated as the 'balancing item' – as the most
expensive form of energy taking up the remainder of our energy
requirements. Underlying this forecast is the assumption that oil
prices are maintained at present levels in real terms and that the
price of North Sea oil will be determined by world prices dictated
by Middle East producers; i.e., that the British government will
not attempt to force the price of North Sea oil below the world
price of crude.

TABLE 5.2 ESTIMATED UK ENERGY CONSUMPTION, 1973–80
(m.t.o.e.)

	Nuclear and hydroelectricity	Natural gas	Coal	Petroleum	Total
1973*	7	23	78	94	202
1974	8	31	76	90	205
1975	9	35	75	90	209
1976	9	39	74	90	212
1977	10	44	73	88	215
1978	11	57**	72	78	218
1979	12	57**	71	82	222
1980	13	57**	71	84	225

Conversion factor: 1 ton of petroleum = 1.7 tons of oil
*1973 figures estimated from *Monthly Digest of Statistics*
**Including natural gas from Norwegian sector of Frigg

Our estimates present a picture of likely future energy requirements which is radically different from that put forward by previous forecasts. The 1972 National Institute 'lower' forecast for 1980 suggested that total energy requirements would amount to 260 m.t.o.e.[7] Our more conservative estimate springs from the expectation of a lower growth rate of GDP (2½ per cent compared with 3½ per cent for the National Institute) and from the view that the higher price of energy will induce improvements in energy conservation and utilisation (hence we assume an energy coefficient declining from 0.7 to 0.6 compared with the National Institute coefficient of 0.8).[8] If these assumptions prove to be in error it appears more likely that they will result in an *overestimate* of energy requirements. Thus, the rate of economic growth may fall short even of the modest target we have chosen and the rise in the price of energy may lead to much greater efficiency in energy utilisation.[9] If so, the eventual improvement in the balance of payments will be greater than that we arrive at in table 5.5.

We have made allowance for recent natural gas discoveries and have assumed that the fall in coal output will continue more slowly than in the past. These factors, and the predicted lower growth of total energy requirements, seem likely to produce a major shift in the pattern of energy consumption. Thus table 5.2 predicts that higher oil prices will bring an abrupt end to the rapid rise in oil consumption of the 1960s and early 1970s and suggests that oil consumption will fall as natural gas takes a higher share of the total energy market.[10]

All this points to a much lower oil energy requirement than the pre-1973 National Institute forecast – only 84 million tons by 1980 as against 133 million tons. In turn this means, given the low price elasticity of the demand of the transport industries for petroleum as a fuel, that there will be extensive substitution of natural gas for oil in domestic heating, industrial use and in power stations.[11] Alternatively, much the same effect on the balance of payments would result if liquified natural gas was exported and its place in the home market was taken by imported crude.

5.3   THE TRADE IN OIL AND THE CURRENT ACCOUNT

The assessments of the balance of payments effects turn on comparing the hypothetical situation that would have occurred if there had been *no* North Sea oil and gas discoveries (North Sea 'off') with the likely situation *given* North Sea oil and gas discoveries (North Sea 'on'). Following our previous practice we shall concentrate on the oil and gas discoveries made in the middle and north North Sea (the Scottish sector), ignoring the earier discoveries in the southern sector whose effects over the past period are much more minor and can be more easily calculated.

We will adopt the simplifying assumption that all North Sea oil is exported and all North Sea gas is consumed in the United Kingdom,[12] although as far as the balance of payments is concerned it makes little difference if we assume that North Sea oil is substituted for imported crude.[13] Hence we take the volume of exports to be equivalent to likely output as previously estimated in table 5.1. Crude oil imports to meet domestic consumption in the North Sea oil 'on' situation can be derived by taking oil consumption requirements as shown in table 5.2 and adding (1) an allowance for stock-building, oil for ships' bunkers, refinery losses and other uses which we take as 15 million tons in 1974 rising by 1 million tons annually to 21 million tons in 1980 and (2) the oil equivalent of the assumed import of natural gas from the Norwegian sector of the Frigg field. The resultant estimate of required oil imports *given* UK discoveries of oil and natural gas in the North Sea is shown in table 5.3.

In the North Sea 'off' situation there are no exports of oil. To calculate import requirements we make the same assumptions regarding total energy requirements and hydroelectricity, nuclear power and coal output, take natural gas from the southern sector of the North Sea to amount to a plateau of 4000 m.c.f.d. by 1975, and assume that the remainder of energy requirements would have been met by imported crude. To this we then add, as above, an allowance for stock-building, oil for ships' bunkers, refinery losses and other uses.

The end result of this process, summarised in table 5.3, shows that in the North Sea oil 'on' situation the UK will no longer be a significant net importer of oil by 1978. By 1980 net exports are estimated to amount to 43 million tons compared with net imports of 127 million tons in the North Sea 'off' situation. The effect of this on the balance of trade in crude oil depends crucially on the price assumptions that are now made. We have valued crude oil exports and imports at 1974 prices; i.e. we have assumed no change in oil prices over the period up to 1980. Further, as light, low-sulphur crude generally enjoys a market premium over heavier crudes, we have assumed that North Sea oil will command a higher price than imported crude. Accordingly we have valued imported crude at $10 per barrel (£32 per ton)[14] and North Sea crude landed in the UK at $11 per barrel. Export earnings will be higher than this as some of the crude will be exported in British tankers and shipping earnings will be remitted to the UK. Hence we have valued exports at $11.50 per barrel (£37 per ton). The resultant visible balance of trade can now be calculated for the North Sea oil 'off' and the North Sea oil 'on' situations (see table 5.3).

The turnabout in the net balance of trade in crude oil is quite staggering. In the North Sea oil 'off' situation it is estimated that the value of crude oil imports in 1980 would be £4064m, a sum equal to six times the total deficit on current and capital account in the 1964 balance of payments crisis. In the North Sea oil 'on' situation it is suggested that the value of crude oil exports would exceed the value of crude imports by £2166m, a turnaround of £6230m. Without the discoveries made in the middle and north North Sea the total oil deficit resulting from crude oil imports would have amounted to £25,824m over 1974–80, compared with a much smaller deficit of £6432m in the North Sea oil 'on' situation. It is possible to exaggerate the economic importance of North Sea oil and gas, but these calculations clearly illustrate that the exaggeration, while still possible, is difficult.

TABLE 5.3 ESTIMATED EXPORTS AND IMPORTS OF CRUDE OIL AND
NET BALANCE OF TRADE AT 1974 PRICES: NORTH SEA 'OFF'
AND NORTH SEA 'ON'

	1974	1975	1976	1977	1978	1979	1980
North Sea 'off'							
Volume							
Exports (million tons)	—	—	—	—	—	—	—
Imports (million tons)	105	106	111	115	119	124	127
Value							
Imports (£m)	3360	3392	3553	3680	3808	3968	4064
Net balance (£m)	−3360	−3392	−3552	−3680	−3808	−3968	−4064
North Sea 'On'							
Volume							
Exports (million tons)		6	28	60	91	137	158
Imports (million tons)	105	106	107	106	106	111	115
Value							
Exports (£m)		222	1036	2220	3367	5069	5846
Imports (£m)	3360	3392	3424	3392	3392	3552	3680
Net balance (£m)	−3360	−3170	−2388	−1172	− 25	+1517	+2166

A few illustrations might underline the point. By 1980 the estimated
value of crude oil exports (at 1974 prices) will be equivalent to slightly
more than one-half the value of all UK exports and 56 per cent of
UK manufacturing exports at 1973 prices, one-fifth greater than
1973 overseas earnings from all machinery and transport equipment
sales and no less than five times the value of motor vehicle exports.
By the late 1970s the recent rise in oil prices will, if maintained,
be operating in our favour and will remove the need to finance a
huge oil deficit. More than that, by the late 1970s the UK can
if it chooses be a net exporter of oil and this will establish a strong
balance of payments position. However, we are still anticipating the
final balance of payments position because thus far we have considered
only the impact on the current account as a result of the trade in
crude oil. To arrive at the overall effect on the balance of payments
some further calculation is necessary.

5.4   PROFITS, ROYALTIES AND TAXATION

Table 5.3 summaries the transactions that would arise directly from the trading in oil in a North Sea oil 'off' situation. In the North Sea oil 'on' situation further adjustments are necessary before we can make a final assessment of likely balance of payments developments. However, before proceeding to this directly we have to consider how the revenue from oil production will be shared between company profits, royalties and government taxation. Ideally the calculations should be based on a detailed knowledge of the cost structures and future revenue of the different North Sea oil fields and the nature of the taxation regime that will be implemented. Such knowledge is simply not available. In chapter 2 we provided details of the likely capital expenditure and operating costs of four fields – Argyll, Auk, Forties and Piper – which only serve to demonstrate that there is no such thing as a 'typical' or 'average' North Sea field which we could use as a bench mark. Nor can we know what fields might be discovered and in production by 1980. However, for lack of anything better we have to start with our existing information and attempt some view of the likely division of sales revenue between company profits on the one hand and royalties and tax on the other.

The normal method of proceeding has been to attempt to break down the selling price of an average barrel of oil into capital and operating costs, net profit, royalties and tax take per barrel. A simple accounting estimate of the cost per barrel has usually been obtained by dividing capital expenditure and operating costs by total output over the expected life of the field. Applying this method to the figures shown in chapter 2 (appendix 2.1) would yield a capital cost per barrel as follows: Argyll, $0.16; Auk, $0.88; Piper $0.58 and Forties, $0.95. Operating costs calculated by the same method would amount to Argyll, $2.10; Auk, $1.55; Piper, $0.54 and Forties, $0.93.

This mechanical computation, while it provides some perspective,[15] is hardly an accurate representation of the factors that determine profitability and commercial judgement as these depend not only on costs per unit of output but also on the profile and timing of anticipated expenditure and revenue. When this is taken into account projects like Piper and Forties, which are heavily 'front-loaded' (large capital expenditure early in life and anticipated sales revenue spread over a long forward period) may be less 'profitable' than projects such as Argyll or Auk, which, while they have higher capital and operating costs per unit of output, yield revenues early in their life.

In view of these difficulties, which will certainly affect commercial judgement, we must abandon the simple accounting cost concept as it will produce quite misleading results. Instead, we can proceed by setting a target internal rate of return on capital investment and then calculate the net return per barrel that will be required to yield that rate of return. We take the target rate of return as 25 per cent for the reasons previously advanced,[16] and on the basis of the figures in appendix 2.1 we can calculate that the following net profit per barrel (after royalties and tax) over the life of the project will be necessary to obtain this rate of return on the separate fields: Argyll, $3.15; Auk, $4.03; Piper, $2.91; Forties, $4.81.

These figures are indicative only and they indicate yet again the difficulty of any generalisation. Yet we must still attempt it if we are to obtain any estimate of the likely benefits that might accrue from North Sea oil. What we have attempted to do is to try and indicate the limits to a taxation regime which would still allow a reasonable rate of return on capital investment such as to encourage continued exploration and production of North eSa oil. On the basis of the above calculations it would seem that companies investing in the North Sea might require an 'average' rate of return per barrel (*after* all royalties and taxation, but *before* operating costs are accounted for) of $4. With a landed price of North Sea crude of $11 this would leave a total government take of $7 per barrel, of which $1.375 (12½ per cent of total sales revenue) would be royalties and the remainder, $5.625, government taxation.

Before applying these figures to the estimated output of oil shown in table 3.4 (p. 64), one further qualification is necessary. Capital investment in North Sea operations is subject, like other similar forms of investment, to 'free depreciation' provisions: that is, the cost of investment can be offset against profits so that no tax liability is incurred until the capital value of the asset is written down. Hence, while royalty payments are made on all oil production, most oil fields will pay little or no corporation tax in the very early stages of their productive life. We have calculated that with existing royalties and capital allowances there would be little tax yield for either the Forties or Piper fields in the first two years of production. After two years corporation tax would apply, but at less than the full rate. Auk and Argyll would both earn taxable profits in the second year of production, and with Argyll the full rate of tax would apply. The generalisation we have adopted from this is that the government will receive only royalty payments ($1.375 per barrel) on the first two years' production from any oil field. Subsequently, we apply the computed average company revenue, tax take and royalty payment per barrel as derived previously.

We would emphasise that our assumptions are likely to yield a conservative estimate of the income accruing to the government through royalties and taxation. This is because we have assumed a generous target, an internal rate of return on capital invested at 25 per cent. Indeed, the further assumptions we have made, that the company should obtain a net return after royalties and tax of $4 per barrel and that free depreciation provisions would apply, would probably give an internal rate of return on most oil fields significantly in excess of 25 per cent. Nonetheless, as table 5.4 clearly demonstrates, even these conservative assumptions yield a huge government income in the form of royalties and taxation.

TABLE 5.4 COMPANY PROFITS, ROYALTIES AND TAXATION, 1975–80
(£m)

		1975	1976	1977	1978	1979	1980
1.	Total sales revenue* of which:	209	986	2095	3186	4818	5534
2.	Company revenue**	183	863	1186	1739	2355	2191
3.	Royalties	26	123	262	398	602	692
4.	Tax	0	0	647	1049	1861	2651
5.	Total government take ((3)+(4))	26	123	909	1447	2463	3343

*Valued at $11 per barrel, compared with $11.50 per barrel (including shipping profits) used to value exports in table 5.3.
**Revenue *after* royalties and corporation tax, but *before* operating costs.

Given free depreciation, government revenue over 1975–6 is likely to be modest as only royalties will apply (on our simplifying assumptions). From 1977 corporation tax begins to take effect and government revenue will mount rapidly. By 1980 the great bulk of all

North Sea production will be subject to tax at the full rate and it is estimated that total government revenue will amount to £3343m, no less than 41 per cent of all taxes on incomes and 46 per cent of all taxes on expenditure in 1972. Subsequently, as output increases beyond 150 million tons a year (we estimate that a level of 180 million tons a year can be sustained for a considerable period from 1982 onwards)[17] government revenue is expected to increase to reach some £4000m annually.

5.5 THE 'OIL' BALANCE OF PAYMENTS

The transactions that would arise directly from UK trade in crude oil in the North Sea oil 'off' situation are summarised by table 5.3.[18] In the North Sea oil 'on' situation certain further allowances are to be made before we can arrive at an assessment of the 'oil' balance of payments. Four further factors are important:

1. the extent to which capital and operating expenditure on North Sea oil and gas is financed by inward investment from overseas;

2. the import content of exploration expenditure and capital expenditure on North Sea oil development;[19]

3. the import content of operating expenditures; and

4. profits and interest remitted overseas as a consequence of capital investment under (1) above.

In the light of our previous discussion we can deal quickly with the first factor. Table 4.3 (p. 75) shows estimated capital expenditure over 1973–80 on exploration and oil developments. To date some 70 per cent of this expenditure has been financed from overseas. Even given the announced intention for government participation in the oil developments it appears unlikely to influence direct inward capital investment in the immediate future and we have therefore assumed that this proportion will be maintained throughout the 1970s.

It is extremely difficult to form any clear view of likely trends in the second factor as they turn on the extent to which UK industry can adapt to the requirements of the new market and on the ability of the UK government to adopt the policy instruments which will maximise UK investment. The IMEG report estimated that UK industry's share of the off-shore North Sea market might expand from 25–30 per cent in the early 1970s to 35–40 per cent in the late 1970s because of the advantages of proximity to North Sea activities and the accumulation of relevant expertise for off-shore exploration and development. Given the adoption of appropriate policies IMEG considered that the share of British industry might expand to 70 per cent by the late 1970s, but past experience does not suggest that this optimistic target can be reached. Our calculations suggest that at the present time the UK share of the off-shore market in the UK North Sea is still less than 35 per cent.[20] We have suggested various policies that might increase the UK's market share and on the assumption that these are adopted and prove effective we have allowed for a steady increase in the UK share of the off-shore market as follows: 1974, 35 per cent; 1975, 37 per cent; 1976, 39 per cent; 1977, 41 per cent; 1978, 43 per cent; 1979, 45 per cent; 1980, 47 per cent.[21]

The regularity of this increase is quite artificial as, given the very 'lumpy' nature of North Sea investment, the UK share of the market is likely to fluctuate quite sharply from year to year. The projected increase in the share might be regarded as pessimistic, but while this has quite important implications for the direct employment created from North Sea activities it is of only minor significance compared with the huge balance of payments benefits that will arise from the trade in crude oil. Thus, even if we assumed that all North Sea oil equipment and services were supplied by UK firms, the additional benefit that this would bring to the balance of payments would still be of only minor significance relative to the receipts from the export of crude oil. Our own view would be that the projected increase in the UK share of the off-shore market is likely to prove optimistic rather than pessimistic.

We have previously calculated that operating expenditures per barrel of production will be as follows: Argyll, $2.10; Auk, $1.55; Piper, $0.54 and Forties, $0.93.[22] Forties and Piper are likely to be more typical of North Sea conditions than Argyll and Auk and as the former are much larger fields we have adopted a convention based on their likely experience: this is that operating expenditure will amount to $1 a barrel. We assume that 20 per cent of these expenditures will be supplied from overseas as the import content will certainly be much smaller than for exploration and capital expenditures.[23]

Applying this convention to the estimated production figures of table 3.4, we obtain total operating expenditure and this is then deducted from company revenue in table 5.4 to give company profits after revenue, tax and operating expenditures. Company revenue and its distribution between operating expenditures and profits is given in table 5.5.

TABLE 5.5 COMPANY REVENUE AND ITS DISTRIBUTION BETWEEN
OPERATING EXPENDITURES AND PROFITS
(£m)

	1975	1976	1977	1978	1979	1980
Company revenue* *of which:*	183	863	1186	1739	2355	2191
Operating expenditures	19	90	190	290	438	503
Profits	164	773	996	1449	1917	1688

**After* royalties and tax

Finally, we require some assumption about that share of profits and interest earned on capital invested in the UK North Sea which will be remitted overseas. We have worked on the supposition that foreign direct investment will finance 70 per cent of capital expenditure, but the overseas share of profits is likely to be less than this as UK companies, partly as a consequence of favourable treatment during the licensing rounds, have been particularly successful in exploration. With this in mind we have adopted, rather arbitrarily, the assumption of a 45/55 division between profits retained in the UK and those remitted overseas.[24]

The results of these different calculations are summarised in table 5.6.

Over 1974–77 the inflow of capital investment for the North Sea outweighs remitted profits and interest with the result that the balance of current and capital accounts is more favourable than the net balance on visible trade in crude oil (compare row 5 with the final row of table 5.6). Thus inward investment in the North Sea will slightly reduce the oil deficit over these crucial years. As profits and interest remitted overseas rise, the balance of current and capital accounts becomes, over 1978–80, less favourable than the current balance on visible trade in crude oil. Our estimates almost certainly overstate the extent to which this will occur for, as previously observed, we have based our calculations on a very liberal regime which is, *vis-à-vis* the oil companies, probably generous to a fault. A more realistic regime would be likely to yield a surplus on current and

TABLE 5.6 'OIL' BALANCE OF PAYMENTS, 1974–80

	1974	1975	1976	1977	1978	1979	1980
1. Exports (oil)	*	222	1036	2220	3367	5069	5846
2. Imports (oil)	3360	3392	3424	3392	3392	3552	3680
3. Imports of equipment and services for UK sector of North Sea	488	646	732	664	584	495	411
4. Import content of operating expenditures	*	4	18	38	58	88	101
5. Visible Trade Balance (1)−[(2)+(3)+(4)]	−3848	−3820	−3138	−1874	− 667	+ 934	+1654
6. Invisibles: profit and interest remitted overseas	0	90	425	548	797	1054	928
7. Current Balance (1)− [(2)+(3)+(4)+(6)]	−3848	−3910	−3563	−2422	−1464	− 120	+ 726
8. Inward investment in UK North Sea	525	718	840	788	718	630	543
9. Balance of current & capital accounts [(1)+(8)]−[(2)+(3)+(4)+(6)]	−3323	−3192	−2723	−1634	− 746	+ 510	+1269

*negligible

capital account of some £1500m by 1980 compared with the estimated surplus of £1269m. Yet, even on the basis of our conservative assumptions, the transformation achieved by North Sea oil and gas discoveries is remarkable. In the North Sea 'off' situation we calculated an oil deficit of £25,824m over 1974–80 (see table 5.3). This compares with an estimated deficit on the 'oil' balance of payments of £9839m in the North Sea oil 'on' situation shown in table 5.6. After 1980 we would expect the surplus to continue to increase to reach a level in excess of £2000m annually, as there is every reason to suppose that oil production can be maintained in the region of 180 million tons a year over a considerable period.

5.6   CONCLUSIONS

The economic and strategic importance of North Sea oil and gas lies in the fact that it will eventually relieve the need for large-scale borrowing to finance the huge oil deficit caused by the quadrupling of crude oil prices. More than this, it seems likely that by the late 1970s North Sea oil production will be in excess of internal consumption needs so that the UK can become, if it so chooses, a net exporter of oil. We believe that this will happen. Whatever present statements are made to the contrary, it will be difficult for any British government to resist the temptation to export sufficient oil to service, and possibly redeem, the international debt that will be incurred in servicing the oil deficit over the period to 1978. The existence of North Sea oil does not therefore obviate the need for borrowing in the short term, or the need to perceive correctly the function of such borrowing.[2b] But it will make it easier to secure the necessary loans as it offers the prospect (as long as oil prices remain at their present levels) of a substantial balance of payments surplus by the late 1970s.

Yet, while North Sea oil offers the prospect of eventual relief from the crippling size of the current oil deficit, the relief is far from immediate. The 'oil' balance of payments will remain in deficit, although the deficit will diminish, for a number of years. Over all this hangs the prospect that oil prices might break by the late 1970s when the UK begins to emerge as a potentially significant net exporter. If so, the UK would have borrowed against high current oil prices only to find that its ability to service that debt was diminished by falling export prices for its own crude. For example, a price for imported crude of $7 a barrel and for North Sea crude of $8 a barrel would produce a surplus on the 'oil' balance of payments of only some £750m in 1980. This simply stresses the need to direct energy conservation policies towards saving on imported crude in the short and medium term, while attempting to minimise any further 'slippage' in North Sea oil production.

Leaving this aside for the moment and assuming the continuation of present crude oil prices (in real terms), we can see that the chief immediate beneficiary of this improved balance of payments will be the Exchequer, and this must follow so long as North Sea oil is sold at world prices. In theory it would be possible for the government to force down the price of North Sea oil so that the consumer benefited directly through the reduced price of energy. In practice, effective price regulation would require extensive direct control over production and distribution and it is difficult to see how this could stop short of outright nationalisation. This would be administratively inefficient and economically disastrous, as the production of North Sea oil is, as we have seen, heavily dependent on foreign capital and foreign expertise. Moreover, if the price of crude was forced down significantly below world prices it would necessarily mean that some North Sea fields, which would be economic to exploit at world prices, would prove uneconomic.[26]

It also follows, if North Sea oil is sold at world prices, that a special taxation regime is necessary for the industry. In a competitive market no special treatment would be necessary as competition between producers would ensure that excess profits were not earned. However, the price of crude is not set in a competitive market; if it were it would not be standing at present levels. The price of crude is set in a market where, as we have seen in chapter 1, there are strong monopoly elements. In consequence, the price of oil will be above its minimum supply price and the difference must accrue either as high company profits, or as government taxation and/or royalties. As North Sea oil accounts for only a small part of world reserves and cannot have any substantial effect on world prices, it follows that the major direct beneficiary will be the UK government. However, while the consumer will continue to pay high prices for imported crude from the Middle East, foreign purchasers of North Sea crude will pay high prices, much of which will accrue under any system of taxation to the British government. If Britain *is* a net exporter of oil by the late 1970s, which our calculations suggest is indeed possible, then the high price of crude will from that date work to her advantage.

This prospect may not be very alluring to the ubiquitous 'man in the street', but it gives rise to other indirect benefits of considerable significance. First of all, we should note that the UK will continue to import substantial quantities of heavy crude from the Middle East. The effect of the increase in the price of crude is then deflationary. In the short run, to offset this, the government must attempt to borrow to finance the oil deficit on the balance of payments, thus maintaining the level of demand and preventing an increase in unemployment. As North Sea oil production grows it is expected that the balance of payments position will improve, but as much of the sales revenue from North Sea oil will accrue to the Exchequer in royalties or taxation, the end result will still be significantly deflationary unless the government increases expenditure or reduces taxation in other directions.

We should underline the above argument because it can be the source of considerable confusion. It is quite simply bad economics to suppose that current borrowing to finance the oil deficit will 'mortgage' North Sea oil so that nothing will be left over for regional development, or for other purposes. It is true that Britain may wish to become a net exporter of crude in order to service borrowing to cover the oil deficit over 1974–8.[27] However, after 1977, it will still be necessary, *unless the government wishes deliberately to engineer a significant departure from full employment,* to take measures to offset the deflationary effect of the higher price of imported crude. In other words, unless the government wishes to reduce demand and deliberately create unemployment, it cannot use the major part of the oil revenues to create a budgetary surplus. Instead it must take measures to stimulate demand by increasing expenditure, or reducing taxation. Indeed, the substantial revenue of North Sea oil gives the government considerable freedom of action – a point of considerable importance when we come to our discussion in the final chapter.

The improvement in the balance of payments position as a consequence of North Sea oil and gas discoveries will have a favourable effect on income and employment as it will allow much more expansionary demand management policies than would otherwise have proved practicable. This conclusion can be disputed, as in principle macroeconomic policy could aim at maintaining the same level of demand, and hence the same level of incomes and employment, irrespective of any deficit on the current account of the balance of payments. In a North Sea 'off' situation this would prove manageable as long as the government was prepared to accept, and was able to finance, the huge deficit on the balance of payments. It is always *possible* to accept this analysis – all that is required is excessive optimism and considerable ignorance of Britain's recent economic history. Lacking either, we must conclude that both the past record of economic management and the lamentable growth performance of the British economy, indicate that in the absence of North Sea oil demand management would necessarily have been more restrictive with adverse effects on employment and economic growth. The existence of North Sea oil will allow any given balance of payments target to be achieved at a higher level of employment, income and output.

There is one further important effect which we analyse in detail in the following chapter. This is the most obvious impact – the impact on employment and incomes resulting directly from North Sea oil If the North Sea oil and gas discoveries had been made in a region high unemployment, Scotland,[28] this will permit a higher *national* level of income and employment than would otherwise be possible. If the North Sea oil and gas discoveries had been made of a region with full employment then much of the increased employment in oil-related activities (assuming limited labour mobility between regions) would have been at the expense of employment and output in other sectors of the local economy and would therefore have had a high opportunity cost. In Scotland, with a level of unemployment significantly above the British average, it should be possible to bring into employment people who would otherwise be unemployed, or engaged in low-productivity, low income employment. In other words, the opportunity cost of these resources will be low.

This over-simplifies a very difficult problem as the major employment impacts of North Sea activities are being felt outside the main centres of population in Scotland, often in small communities with restricted labour catchment areas. Hence, considerable labour mobility may still be necessary if the increase in the national level of employment and income is to be secured. Nonetheless, it is still true that North Sea oil activities will help to provide a better regional 'balance' and hence a higher macroeconomic level of output and employment than could otherwise be possible. It will still be true that the direct employment and income effects of North Sea oil and gas activities will be small relative to the improvement which will result in the balance of payments, or relative to the revenue accruing to the government in royalties and taxes. This is because the resource cost of North Sea oil and gas is low; i.e., its value added is high relative to the resources necessary for production and exploitation. Direct employment and income creation will be a function of the capital and operating expenditures necessary to obtain North Sea oil and gas. Although the capital investment required may appear high by absolute standards, it is still small relative to the estimated revenue from oil production, or the estimated government revenue. Inspection of tables 4.3, 5.4 and 5.6 will show that over 1974–80 the capital investment in North Sea activities is small relative to government revenue and foreign exchange receipts from exports, and *this discrepancy will be much more marked after 1980.*

Hence, even if we assumed that all equipment and services for North Sea activities were produced by UK industry, the direct employment and income effects would still be relatively restricted. Of course, such an assumption is quite unrealistic as the equipment and services used in the UK sector of the North Sea have now, and will continue to have in the future, a substantial import content. This may be considered an unduly pessimistic view, but we believe it to be justified. However, given its importance, it requires more detailed investigation. This is undertaken in the following chapter, which looks at the direct employment effects with particular reference to Scotland, which is likely to receive the main brunt and the main direct benefits of North Sea oil and its associated activities.

NOTES TO CHAPTER 5

1. G. L. Reid, K. Allen and D. J. Harris *The Nationalised Fuel Industries* London, Heinemann (1973) p. 119.
2. *Production of Reserves of Oil and Gas in the U.K.* London, Department of Energy (1974).
3. Natural gas from the Norwegian sector of the Frigg field, which will be piped to the United Kingdom, must be treated separately as it represents imports to the UK.
4. *Plan for Coal* London, HMSO for National Coal Board (1950).
5. See M. V. Posner *Fuel Policy: A Study in Applied Economics* London, Macmillan (1973).
6. This prediction runs counter to the National Coal Board's view that coal output can be expanded to 150 million tons by 1985, but is consistent with the detailed assessment by G. Armstrong 'Coal and the Energy Crisis'

The Economic Significance of North Sea Oil   109

British Association for the Advancement of Science (1974).

7. We make the comparison with a pre-1973 forecast simply to illustrate how the view of the future has changed as a consequence of the increase in crude oil prices. We do not, of course, imply that the 1972 forecast reflects the present view of the National Institute.

8. For the National Institute estimates see *National Institute Economic Review* (November 1972).

9. One detailed post-1973 forecast assumes much lower energy coefficients than those we have adopted and forecasts energy requirements of 215 m.t.o.e. by 1980 compared with our estimate of 225 m.t.o.e. See the *Petroleum Economist* (June 1974) pp. 221–3.

10. Indeed, this process seems already to be under way. If allowance is made for the three-day week it appears that UK petroleum consumption in 1974 was 7–8 per cent below 1973 levels: *National Institute Economic Review* (August 1974) p. 25.

11. The predicted level and distribution of energy requirements shown in table 5.2 are very similar to the estimates arrived at independently by Armstrong op. cit.

12. Only some 17 per cent of current UK oil consumption is of lighter crudes similar in quality to North Sea oil discoveries.

13. It is not correct to say that it makes no difference, as the price of North Sea oil crude is likely to be slightly above the price of imported crude from the Middle East.

14. This is below the November 1974 c.i.f. price of imported crude, but some allowance should be made for UK shipping earnings and the remitted profits of UK oil majors in calculating balance of payments effects.

15. Capital expenditure and operating costs are a guide to likely employment creation from North Sea activities. See p. 169.

16. See p. 65.

17. See p. 65.

18. Ideally, to complete the calculations we should make some allowance for the profits remitted by UK oil majors from overseas operations and for the foreign exchange earnings from the crude surplus to UK consumption requirements which is transhipped or refined and then exported from British ports. However, neither of these transactions is directly concerned with the major trade in imported crude and they have rather small balance of payments implications.

19. We, of course, exclude expenditure on gas development. Ideally, we should also exclude expenditure on gas exploration, but the sum involved is relatively small and it is conceptually and practically difficult to distinguish from expenditure on oil exploration.

20. See p. 83.

21. The remainder represents imports and this has been applied to estimated expenditure on exploration and oil developments as shown in table 4.3. Expenditure on gas development has, of course, been ignored.

22. See p. 97.

23. A relatively high proportion of operating expenditures consists of wage costs, while the development phase is much more capital-intensive with a correspondingly higher import content.

24. Of course, some of these profits will be retained and reinvested rather than automatically remitted overseas. However, the effect of this, as far as the balance of payments is concerned, might be largely offset by a reduced capital inflow. For this reason we have ignored the complication.

25. The function of such borrowing is to maintain aggregate demand and

in particular to maintain the level of investment. See W. M. Corden and P. Oppenheimer *Basic Implications of the Rise in Oil Prices* London, Trade Policy Research Centre (1974).

26. This is precisely what has happened in the case of gas from the southern North Sea. We must hope that the lesson has been absorbed.

27. Our calculations in table 5.6 suggest that the total oil deficit over 1974–8 will amount to £11,618m and the cost of servicing this debt is likely to be in the region of £1200m annually.

28. And otherwise largely in north-east England.

CHAPTER 6

Employment Creation in Scotland

The Country in its time may not produce gold or silver or wine or oil, but if it produce industry all these things will be added unto it. (Extract from letter to Countess of Sutherland from Patrick and William Young, 1809[1])

6.1 INTRODUCTION

The major benefits of the oil and gas discoveries will be the increases in government revenue and company profits and the savings on the balance of payments. An outstanding feature of the oil industry is the substantial difference between production costs and selling prices – in other words, the industry generates a very high added value, and much of the political debate concerning the North Sea has concentrated on the distribution of this added value. Chapter 4, however, dealt with the opportunities for capital investment, and a related benefit is the employment created by the developments, particularly as much of the employment creation is taking place in Scotland and north-east England which have had lengthy histories of unemployment. Indeed, in Scotland the major direct effects are the increase in employment and the consequential increase in income. This chapter deals with these employment effects within Scotland, both from the point of view of what has happened to date and what is likely to happen over the next decade, and the two remaining chapters of the book consider the implications of these effects.

The main point to bear in mind is that, although Scotland is an industrialised country, the North Sea discoveries have been made off the more rural and sparsely populated areas. Consequently there has arisen a range of problems related to the marrying of the oil industry

111

with the traditional economic and social structures of these areas. In the main the crucial factors are the scale of the developments relative to the size of the communities affected and the lifetimes of the various developments.

By way of introduction the chapter presents a brief description of the types of employment created by North Sea activities and then considers the present distribution within Scotland. The effects on three areas in the North – the North East, East Ross and Shetland/ Orkney – are examined in more detail followed by discussion of likely future trends. The concluding section draws together the main implications for the North of Scotland and the rest of the economy.

6.2　A Taxonomy of Employment Creation

For ease of understanding, it is useful to revert to the three-phase classification of activity used earlier: the exploration, manufacturing and production phases. In addition, it is helpful to distinguish a fourth phase – that of (temporary) construction activity, i.e. the construction of refineries and terminals, pipe-laying etc. Each of these phases will have secondary effects on incomes and employment through the operation of the multiplier process. The initial direct increases in employment will generate a rise in consumption expenditure and hence increases in income and employment in other sectors of the economy (e.g. shops, hotels, schools). This point is dealt with in more detail below.

The exploration phase covers surveying (magnetic, seismic, geological) and drilling and includes the provision of marine and air transport and the wide range of back-up facilities, materials and equipment. Most of the employment created is either on the rigs or in the service bases supplying the rigs. These on-shore bases are usually located as close as possible to where exploration activity is under way and, as most of the activity is in Scottish waters, the bases have been located on the east coast of Scotland – principally Aberdeen, with smaller bases at Peterhead, Dundee, Montrose and Leith on the mainland and in Orkney and Shetland. Great Yarmouth established itself as the main base for the southern North Sea and still performs this function for exploration and production in that area. With a few wells being drilled in the Irish Sea (stealthily renamed the Celtic Sea by official bodies) one or two small service bases are being set up there.

Precise measurement of the employment effects is impossible because of the mobility of the rigs and boats involved and also because many of the crews are foreign and only use the on-shore bases as staging posts when going on or returning from leave. Nevertheless, some generalisations are possible. On average, a rig in the North Sea requires a crew (including contract employees) of approximately 140 (i.e. two crews of 70, one crew being on the rig at any one time)[2] and generates about 120 direct jobs on-shore. These latter will be the employees of the exploration companies, the crews of the supply boats and helicopters and the employees of the various specialist service companies. A simple estimate of employment creation can be derived therefore by multiplying these proxy figures by forecasts of the number of exploration rigs in the United Kingdom sector. Indeed, going back over past data, this appears to be a more reliable method than would be expected – and would normally be used by economists!

On average about 60 of the 260 jobs (140 off-shore, 120 on-shore) per rig will be filled by non-United Kingdom nationals not normally resident here. Most of the jobs will be semi-skilled or unskilled but quite a few will be for highly skilled personnel – petroleum engineers, drilling engineers, tool-pushers, geologists, divers – and the Department of Employment in conjunction with the Petroleum Industry Training Board has been assessing possible shortages in these skills.[3] Problems are unlikely to be serious in the long run, however, because of the ease in attracting personnel from overseas – e.g. from Australia's declining mining industry – and because of the expansion of appropriate courses in universities and technical colleges, but there are signs of shortages in some key skills such as divers and helicopter pilots.

The second phase identified above was the manufacturing phase, which covers the fabrication (or construction) of the various facilities required for exploration and production. The locational characteristics of this phase differ enormously from those of the exploration phase and, indeed, vary widely according to the item being manufactured. Production platforms, for example, have very specific site requirements. On the other hand, some of the equipment is highly mobile – drilling rigs and supply boats, for example – and can be (and is being) built overseas. Furthermore, manufacturers of other items prefer to be located close to large centres of heavy manufacturing and engineering industry, such as west-central Scotland.

In the main, activity and employment in this phase can be related directly to known discoveries and development plans. As most of the employment is embraced by major activities such as platform construction, module construction and pipe-coating, it is relatively easy to estimate once development plans are known. Problems do arise with subcontracting, although there are certain ways of obtaining information on the extent to which this occurs, e.g. by analysing exemptions from the power restrictions during the three-day week early in 1974.[4]

In terms of the types of labour demanded, the steel platform-builders at present employ around 6000 people, a large proportion of whom are welders, riggers and fabricators. With the increasing preference for concrete platforms, labour requirements will fall (on average from 1500 per platform to around 650) and the skill requirements will approximately closely to civil engineering work. Again, the Department of Employment and other official bodies have considered the appropriate sources of labour supply. Undoubtedly, in certain areas and industries there will be problems of labour shortage but the downturn in the construction industry and the national economy in general should alleviate these problems. Increased efforts to divert footloose activity to the more industrialised areas will also reduce the need for geographical labour mobility and infrastructure provision.

Turning to the production phase, the main feature here is that employment levels will be considerably lower than those in the other two phases. There are similarities between the exploration and production phases in that the operation and servicing of rigs and platforms have common aspects. Generally, however, employment on production platforms and related on-shore employment will be less than on exploration rigs. It is difficult to generalise because the crucial variable is the amount of equipment on the platform and this will vary from field to field. The Auk platform, for example, will require only 10 to 15 people while each of the Forties platforms will probably require 80 people (i.e. two crews of 40) once the development drilling has been completed. In addition there will be employment in the landfall terminals and any new refineries or refinery expansions using North Sea oil or gas.

In their own right the landfall terminals employ very few people. BP's Cruden Bay terminal will be almost wholly automated and will employ less than 30 people on a permanent basis, although there will be approximately 60 people employed in the separation plant at

Grangemouth and about 150 in the Firth of Forth tank farm and terminal. The Flotta terminal for the Piper field will only require 90 permanent employees, and a doubling of capacity to take oil from the nearby Claymore field would only increase the labour force to about 120. Further north in the Shetland Isles, the Sullom Voe terminal will be much larger and will employ up to 400 people when it is fully operational. There are additional possibilities related to refining and processing activities but the likelihood of these being established in Scotland on a significant scale is very small, particularly in relation to output levels from the North Sea. Given the capital-intensive nature of refineries, the employment creation from any new refining capacity will be smaller still.

During the last twenty-five years there has been a substantial growth in refining capacity in the UK, roughly in line with the growth in demand for oil and oil products. Refining capacity has increased from 10 million tons per year in 1950 to 150 million tons in 1974. Over this period the rate of growth in demand for oil averaged 9 per cent per year and it was expected that demand would continue to grow steadily and reach approximately 150 million tons per year by 1980. This would have resulted in a healthy balance between demand and refining capacity. Recent rises in the price of crude oil, however, have adversely affected the demand for oil and oil products and demand has fallen sharply. At present prices it appears that UK demand for imported crude in 1980 will be in the region of 127 million tons.[5] Even if no new refineries were built before then, this would imply an overall surplus of capacity of 23 million tons.

On the basis of plans that have been announced to date, it appears that only 30 to 40 per cent of North Sea oil will be refined in the UK with the remainder being exported as crude oil to refineries elsewhere in western Europe, United States etc. The main factor underlying this pattern is the type of oil discovered in the North Sea, which is of the light variety and is more suitable for refining to the lighter products than is Middle East oil. Within the UK there is a fairly small market for these products, however, and the main markets are in the original European Community countries and the United States. As the costs of transporting crude are much less than of transporting those products, the natural wish of the oil companies is to refine crude oil as close as possible to the markets.[6] Consequently none of the oil companies with North Sea fields intends at the present time to build new refining capacity in the UK.

It may well be, of course, that, in the light of the additional benefits to the economy in exporting products rather than crude (product prices are usually £10 to £15 per ton above crude prices), the government will put pressure on the oil companies to refine in the UK a greater proportion of North Sea oil than the companies currently intend. Even if this pressure meets with results, it seems unlikely that there will be a need for additional refining capacity, and if it were refinery expansions and modifications, rather than new refineries, are more likely.

There are bound to be some exceptions to the generalisations as some companies may wish to break into the UK market, but this is likely to happen infrequently. In Scotland, for example, there are six refinery proposals outstanding but only one appears to have a reasonable chance of going ahead and this would employ only 300 to 450 people. In addition, some spin-off development at the existing BP refinery at Grangemouth should create between 200 and 300 new jobs. It has to be recognised that, in the absence of an increase of refining capacity in Scotland, many spin-off opportunities (such as through petro-chemicals) will be lost as virtually all North Sea oil will bypass Scotland in crude tankers.

Employment in the gas-producing phase is likely to be relatively more important as it is unlikely that gas will be liquified and transported in tankers. Gas from the Frigg field and probably future discoveries in northern waters will be piped to St Fergus, where on present plans the terminal and separation complex will employ about 150 people. There are strong possibilities of ancillary developments being established in the immediate vicinity – e.g. feedstock and ammonia plants – and these would employ between 500 and 600 people. Overall, however, the capital-intensive character of the production phase and the peculiar situation regarding the refining of oil in Scotland mean that employment creation during this phase will be much less than the other phases. This point is returned to in the concluding sections of this chapter.

Finally, in parallel with the three main phases of activity, there will be fairly substantial employment in temporary construction work such as pipe-laying and the construction of on-shore terminals. This work tends to be labour-intensive, frequently relying on migrant labour, and is by nature short-term. Most of the projects are typical of civil engineering work in respect of their labour requirements although pipe-laying is a peculiar activity with the foreign lay barges often operating as self-contained units with their own (foreign) crews moving from area to area. At the present time the Forties, Piper and Frigg terminals are all being constructed and activity in this phase will increase as other fields are developed. Inevitably this work is concentrated on the east coast of Scotland.

6.3 DISTRIBUTION WITHIN SCOTLAND

Employment associated with the North Sea developments has already made a substantial contribution to the improvement of the Scottish economy. In this section we provide an estimate of the present level of employment creation and briefly look at its geographical distribution within Scotland. After the more detailed case studies, the likely trends in the future are considered.

Three points need to be made at the outset. Firstly, the estimates deal only with Scotland. At the moment there are between 5000 and 6000 North Sea jobs in England, mainly in the north-east, where platform and module construction is well established and in East Anglia, which is adjacent to the southern North Sea gas discoveries. These numbers should grow over the next few years as more manufacturing orders are placed in England and as the level of exploration activity builds up in areas like the Irish Sea. Secondly, the estimates are subject to the usual disclaimer about margins of error. Obviously those for the present position are reasonably accurate. The Department of Employment collect and publish monthly figures on the numbers employed in companies wholly engaged in North Sea oil and gas activity. There are problems in covering firms for which the North Sea represents only a portion of their output and in tracking down subcontractors, and, where appropriate, we have adjusted the Department of Employment's figures to take account of these deficiencies. It would be impossible, of course, to present a definitive picture. Thirdly, the figures relate wholly to the North Sea and exclude companies like Marathon Manufacturing who are constructing rigs for other off-shore areas. An assessment of the likely scale of employment creation in this sphere of activity is given later in this chapter.

In table 6.1 it is estimated that at the end of 1974 approximately 19,000 jobs had been created in Scotland as a direct result of the North

Sea developments: 6000 in exploration; 8700 in manufacturing; 800 in production; and 3500 in construction. The table breaks down these figures for the various planning regions in the country and it can be seen that at the present time 32 per cent of the employment has been created in the Highlands and Islands and 37 per cent in the North East.

TABLE 6.1 DISTRIBUTION OF NORTH SEA EMPLOYMENT CREATION WITHIN SCOTLAND, 1974

Region	Exploration	Manufac-turing	Production	Construc-tion	Total
Highlands	400	4800	100	700	6000
North East	5000	300	300	1500	7100
Tayside	300	200		100	600
Edinburgh	300	1900	200	1000	3400
Glasgow		1400	200	100	1700
Rest		100		100	200
Scotland	6000	8700	800	3500	19,000

This distribution is very different from the distribution of all other types of employment in Scotland, which is heavily concentrated in central Scotland and the Glasgow region in particular. Quite simply, the geographical distribution of North Sea employment creation does not match up with the geographical distribution of population (and the unemployed), as can be seen from table 6.2. The different fortunes of the main regions are shown graphically in figures 6.1 and 6.2 which clearly indicate the increase in employment and the fall in unemployment in the North East and the Highlands and Islands since the oil and gas developments began. In the Aberdeen area, for example, unemployment rates are at their lowest level since the Second World War.[7] On the other hand, the performance of the Glasgow region and consequently Scotland as a whole has been affected only slightly, and wider considerations, such as the performance of the UK economy and the structural problems of central Scotland, have continued to be much more important. The implications of this for the Scottish economy are discussed in chapters 7 and 8, but the more local implications are dealt with in the following case studies.

TABLE 6.2 REGIONAL DIFFERENCES WITHIN SCOTLAND, 1974

Region	Oil employment	Total employment	Unemployed	Population
	(%)	(%)	(%)	(%)
Highlands	31.6	4.0	4.7	5.4
North East	37.4	7.9	4.8	8.6
Tayside	3.2	8.4	7.5	8.7
Edinburgh	17.9	19.7	16.7	19.7
Glasgow	8.9	48.4	58.8	47.9
Rest	1.1	11.6	7.4	9.7
Scotland	100.0	100.0	100.0	100.0

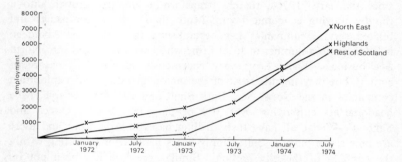

Figure 6.1   North Sea employment creation, 1971–4

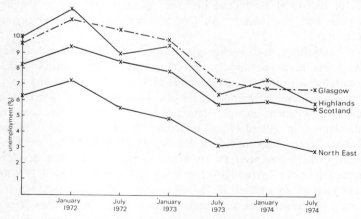

Figure 6.2   Male unemployment percentages, 1971–4

North East Scotland

In 1971 the North East region had a population of 452,000, of whom 182,000 were in the city of Aberdeen. The economy has been dominated by the agriculture, fishing and food-processing industries. With the common decline of employment in the primary sector, throughout the 1950s and 1960s the region suffered substantial population loss through emigration coupled with rural–urban drift from the landward areas to Aberdeen and the larger towns. Since 1970, however, rising prices for primary products have greatly improved the fortunes of the traditional industries and concurrently there has been a rapid expansion in the service sector. It was into this growing economy that the oil companies have steadily moved since the early 1970s, thereby transforming modest growth into a rapidly moving economic boom. More than 200 oil companies and related service companies have established themselves in Aberdeen in addition to a number of local firms who have diversified into North Sea business. These firms vary enormously in size and nature of activity but they include most of the major operating and exploration companies in the North Sea. Both Shell and BP, for example, have headquarters employing over 300 people each; Sedco, Odeco and Santa Fe-Waage all have their drilling bases in Aberdeen; and there are now seven purpose-built supply bases planned or operating within the confines of the city harbour. Similar pressure has been put on industrial space in and around the city and at Dyce airport. Over 300 acres of industrial land have been zoned and taken up in the last two years and proposals for another 500 acres (mainly at Dyce) are now in hand.

An interesting and encouraging aspect of this influx of new activity has been the appearance of companies and bodies largely unconnected with the oil and gas industries. These are over and above the normal indirect, spin-off activities associated with increases in income, employment and population. The most notable has been the establishment in Aberdeen of merchant banks and finance houses (Edward Bates, Bank of Nova Scotia, Finance for Industry etc.). Similarly, there has been an enormous expansion in office developments in the city: in June 1974 existing office space totalled 140,000 square metres; an additional 95,000 square metres had received full planning permission and were under construction; and another 80,000 square metres were at various stages of the planning application process. There are proposals in hand to increase by 80 per cent the number of hotel beds in the area, and the number of scheduled air services to and from Dyce airport have doubled in the last two years, in addition to the upsurge in helicopter and fixed-wing charter flights associated with the exploration rigs.

All these events are indicative of a type of self-generating growth which economists sometimes represent by a 'super multiplier' concept. In other words 'regardless of the initial location advantage ... this build-up becomes self-sustaining because of increasing internal and external economies at these centres'.[8] This is implicit in the belief of the North East of Scotland Development Authority (NESDA) that Aberdeen will consolidate its position as 'Europe's Offshore Capital'[9] in much the same way as Houston has grown in the United States.

At the present time the major constraint on the rapid expansion of the Aberdeen area is the need for substantial infrastructural expenditure to accommodate the growth of employment and population. The best example of this is housing. A recent report prepared by the North East of Scotland Joint Planning Advisory Committee (NESJPAC)[10] forecast a demand for about 16,500 houses over the period 1972–6 to accommodate the growth of the oil industry, indigenous industries and. the area's general needs. This implied an annual rate of housebuilding of around 3500 – twice what has been attained in the period to 1974.

Labour shortages have been widespread throughout the construction industry and certain other sectors of the local economy. Unemployment in the area has been virtually eliminated – the July 1974 percentage figure was 1.4 per cent compared with 4.2 per cent for Scotland and 2.4 per cent for Great Britain as a whole. This has meant that new and expanding firms have had to rely either on immigrant labour or on taking labour from established firms. These latter firms have experienced severe competition for labour and labour losses have been very common. Competition has been exacerbated by the vicious circle that has been created: labour shortages reduce the rate of housebuilding, which in turn aggravates the labour situation. Unfortunately the situation was worsened by the operation of the national incomes policy, which while strictly controlling wages paid by established firms was incapable of controlling wages paid by new firms with the result that oil companies and others found it easy to 'poach' labour (see chapter 7).

Further north, in and around Peterhead, the relative scale of the oil and gas developments has been considerably greater. Whereas Aberdeen has been largely concerned with exploration, administration and a certain amount of manufacturing work (e.g. drilling tools), Peterhead is concerned more with construction activity and, in the long run, with production. The coastline near Peterhead has been identified as one of the best landfalls for pipelines from the North Sea and two such schemes are well under way. Firstly, Forties field oil is to be piped ashore at Cruden Bay, a few miles south of Peterhead, and then transferred by underground pipeline south to the Firth of Forth. This necessitates the construction of a landfall pumping terminal at Cruden Bay, employing a permanent labour force of 25–30. Secondly, gas from the joint UK–Norwegian Frigg field is to be piped ashore at St Fergus, north of Peterhead, and similarly piped to markets in the south. The local impact will be much greater, however, as the facilities required at St Fergus are more extensive and will employ up to 150 people. In fact there will be two terminals – the Total terminal, which will include treatment plant for removing and stabilising liquids, drying the gas and measurement and control, and the British Gas terminal, which will include facilities for blending, compression and pumping. Frigg gas will have a low delivery pressure at St Fergus and will require compressor stations at forty- or fifty-mile intervals along the two pipelines taking the gas south to the national network.

Both the terminals will provide stable and long-term employment in an area which has historically suffered from the pull of Aberdeen. They are on a scale that should cause little disruption locally, although fears have been expressed regarding the short-term situation during the construction of the two terminals and a nearby power station. Peterhead has experienced already a large and volatile construction labour force as the off-shore pipe-laying operations for the Frigg and Forties fields have been conducted from there. During the summer of 1974 this involved a peak labour force of 2000, most of whom were foreign nationals. A concern about the possible repercussions of these floating labour forces led the Church of Scotland to produce a report on Peterhead and the oil industry,[11] following on from a more general report published in 1973.[12] The Peterhead report presented a grim picture of the town itself and the impact of the oil developments, and although it was not generally accepted it raised some pertinent questions about the speed of development and the lack of an overall planning strategy. These points are returned to in chapter 7.

East Ross

The other main area in Scotland in which oil employment has been concentrated to date has been the Moray Firth area in the Highlands. In August 1974 oil-related companies in the area were employing over 5000 people, some 30 per cent of total employment creation in Scotland. In fact, this was almost wholly due to three activities: the two steel platform sites at Nigg (Highlands Fabricators) and Ardersier (McDermott) and the MK–Shand pipe-coating facility at Invergordon. The situation is quite different from that in the Aberdeen area. Instead of a large number of firms engaged in a wide range of activities, the Moray Firth is dominated by three firms undertaking work of an inherently fluctuating nature.

The Moray Firth area embraces the Cromarty and Beauly Firths and the coastal section from Ardgay in the north to Inverness and Ardersier in the south (see frontispiece). For many years this area has been regarded as the most promising in the Highlands in terms of its potential for attracting manufacturing industry to replace employment losses in traditional industries. Since its inception in 1965 the Highlands and Islands Development Board had promoted the Moray Firth area as being one of three major growth centres in the region:

> We will do our utmost to generate major growth points, involving substantial increase in population wherever the natural advantages of the area seem to warrant it; the Moray Firth is unquestionably the most important of these.[13]

In East Ross, the county council lent its weight to these efforts by zoning large areas of land for industrial development and related housing. The efforts came to fruition in the late 1960s with the choice by the British Aluminium Company of Invergordon as the site for its new aluminium smelter. The first signs of impending problems emerged in late 1970. At the peak, a labour force of over 2000 was employed in the construction of the smelter. Many of these were incomers who decided to stay in East Ross (at least temporarily) when the construction phase finished. Hence, as construction work finished, the unemployment figures rose again and reached 14.4 per cent for males in August 1971 – even higher than before the smelter was established. A proposal for a petro-chemical plant never materialised, but the gloom in the area, which had been exacerbated by a post-

ponement of the second-phase recruitment for the smelter, was lifted
in October 1971 when Highlands Fabricators applied for planning
permission for a site at Nigg to build steel production platforms for
the North Sea fields. Highlands Fabricators are a Scottish-registered
company, set up by the Americans Brown and Root and the UK
construction company George Wimpey. The application received
strong support locally and from the county council, who gave approval
in principle in November. The Secretary of State for Scotland made
an Article 8 direction (necessary because the site was not zoned for
industrial use) in December, without holding a public inquiry. In
January 1972 site work began and Highlands Fabrications announced
that they had won the order for the first production platform for the
Forties field.

Highlands Fabricators have virtually mopped up all the local
unemployed in addition to providing work for a large number of
incomers. Great strain has been put on the local economy and public
services in particular, however, in the same way that this occurred in
the North East. The main problem has concerned the numbers of
people employed at Nigg and most of the other problems that have
arisen have really been consequences of this central problem. In their
original application for planning permission Highlands Fabricators
stated that they would employ between 600 and 900 men, the majority
of whom would be recruited locally. In addition, 400 men would be
required for site preparation work. Recruitment began on a large
scale in March 1972 and a training school was established. When a
smaller order was obtained from Amoco the company increased its
expected labour total to 1050. In February 1973 a second BP order
for the Forties field was announced and the labour total was increased
to 1500. With expected completion dates being pushed back steadily,
a revised estimate of 2500 was made in September 1973 but by June
1974 over 3300 people were employed at the Nigg site. Subsequent
to the floating out of the first platform about 1600 people were laid
off in July and August, most of whom had been recruited temporarily
to ensure completion of the platform jacket.

Hence, in only thirty months the labour force had risen from zero to
over 3300. This raised serious problems in regard to labour recruitment and housing provision. At the very time that the county council
and private builders needed more labour to build houses to accommodate Highlands Fabricators employees, Highlands Fabricators
themselves were recruiting heavily from the local construction trades.

In the end, the company solved their problems by chartering two cruise liners which they moored next to the site and used as dormitory accommodation for temporary workers they brought in from outside the area. This brought tremendous increases in trade for some people, not least the pubs and certain other professions, but it also resulted in social problems connected with the existence of a large number of single or unattached males in a quiet, rural area with very few of the social facilities to which they were accustomed.

A related problem has been the substantial fluctuations in the labour force, which eventually resulted in nearly 2000 people being laid off. There has been a similar pattern with the pipe-coaters MK–Shand at Invergordon. In their original application for planning permission the company estimated that they would employ about 150 people over a fifteen-year period. Since they began work in May 1972 they have secured two major contracts – for the Forties and Frigg underwater pipelines – and their labour force has risen to 600, fallen to 50 and risen to over 700 again. Such fluctuations may be easily absorbed by an urban community but they create serious disruptions in an area like East Ross.

Given these problems, it would have been very difficult for any local authority to forecast the requirements for new houses, schools and roads, and to plan accordingly. Nevertheless, a consideration of the role of the county council, its planning committee and planning department throws up questions regarding the ability of small local authorities to cope with large-scale industrial developments of the type generated by the North Sea discoveries. Some of these questions are of general applicability to all local authorities; others would not do justice to the performances of authorities such as Shetland and Orkney. It does appear, however, that Ross and Cromarty county council has added to its own problems.

In recent years the county council has been dominated by a group of pro-industrialists who have alternated between advocating policies of short-term expediency on the one hand and on the other hand the large-scale zoning of land for industrial and housing use well in advance of actual need. The latter policy would undoubtedly have been suited to the particular circumstances in East Ross if the underlying estimates of employment and population growth had been reasonably accurate and realistic. Over the last few years, however, the county council's estimates have been based on a type of 'magic arithmetic' which even the advocates of the new maths system would find confusing. The main basis has been a peculiar transformation of acreage zoned for industry into new jobs and hence population growth, with no regard to the possibilities of attracting firms to occupy the acreage. In September 1972 the county council produced forecasts of employment and population growth of 57,000 and 140,000 respectively over the next twenty years. In April 1973 these were revised upwards to 65,000 new jobs and 160,000 additional people; the related planning proposals included the construction of a new town with a population of up to 100,000. In March 1974 the forecasts were suddenly reduced to 20,000 new jobs and 50,000 additional people.

These enormous changes have been caused by the ignoring of the central issue – what is likely to happen to employment in the area in the next decade or so, and this is largely dependent on oil developments. The forecasting exercise can be divided into two parts: what will happen to employment in existing firms; and what are the possibilities for new firms moving into the area?

Regarding the first question, there are at present only two major oil-related employers in the East Ross area: Highlands Fabricators and MK–Shand. In addition, a dozen or so smaller firms employ about 200 people but most of these are also directly related to the platform and pipe-coating yards. As has been shown above, the labour forces of Highlands Fabricators and MK–Shand have fluctuated substantially in the past and will continue to do so because they are related to individual contracts. Overall, the prospects for steel platforms are not good (see chapter 7) and the peak period for pipe-coating will be in the late 1970s. Highlands Fabricators have invested over £20m in their Nigg site and are capable of diversifying their activities, but it seems reasonable to expect that the normal labour forces of the two companies by the early 1980s will be less than they currently employ.

As regards prospective incomers the position is much less clear. Two other companies have planning permission for sites for production platform (concrete) but they have no orders as yet. If they went ahead simultaneously they would employ around 1500 people between them. Planning permission was given in the heyday of the county council's expansionist policy.

Quite a few other proposals are also in the air and they complicate the issues further. In particular two property development companies – Cromarty Firth Development (a subsidiary of Onshore Investments,

who have been engaged in similar activities in Shetland and Peterhead) and Highland Deephaven – have acquired or taken options on
large acreages of land and have produced wide-ranging proposals
for industrial estates etc. Some of these may materialise, but in the
present situation it is difficult to plan developments sensibly. Again,
the county council has added to its own problems by giving permission to these and other projects without obtaining sufficient information and undertaking the necessary appraisal of likely effects on
the local economy and community.

There are signs of enlightenment, however. The major outstanding
application in the area is that of Cromarty Petroleum for a refinery at
Nigg. This is likely to go to a public inquiry, but in the meantime
the county council has obtained a great deal of information from
consultants, and considerations of environmental costs, stable employment etc. are receiving attention. Nevertheless, a balanced appraisal
of future prospects in the short term almost certainly leads to the
conclusion that future employment and population growth will be on
a much smaller scale than Ross and Cromarty county council is
advocating and anticipating. Present employment levels may well
decline throughout the 1980s as resources are switched from the
manufacturing to the production phase, and in that decade East Ross
and the Moray Firth will have to rely on the non-oil sectors for any
employment and population growth. Such growth will be slow rather
than spectacular.

Shetland and Orkney

These two groups of islands lying off the northern mainland of
Scotland are being affected in the same way as the North East in that
they are largely concerned with the exploration and production phases,
being adjacent to many of the discoveries that have been made.
Shetland (or Zetland) comprises a group of over a hundred islands
with a 1971 population of 17,300. Most people live on the mainland
which contains the only burgh, Lerwick (population 6100), although
there are eighteen other inhabited islands. For many years the staple
industries have been crofting, fishing, fish processing and knitwear.
The Orkney Isles are much closer to the Scottish mainland but have
a similarly sparse population (17,000), most of whom also live on the
mainland which has two small burghs, Kirkwall (4600) and Stromness
(1650). Orkney has a prosperous agricultural industry, some fishing

and two small whisky distilleries.

Most of the fields discovered to date have been in the East Shetland basin. Initially, therefore, a few small supply bases were established in Shetland to service survey vessels and exploration rigs. The geographical isolation of the islands and the small-scale facilities there meant that these have been small, forward bases used mainly for transhipment, and most of the rigs in this area of the North Sea have continued to be supplied wholly or largely from Aberdeen and other ports on the Scottish mainland. Over time, however, it is possible that the size and role of the supply bases in Shetland will increase, particularly as requirements – such as for the servicing of production platforms – have become much clearer.

Of most importance for the Shetlands are the production plans for the fields that have been discovered. Briefly, it is intended to construct a large, multi-user terminal at Sullom Voe in the northern part of the mainland isle. In September 1974 outline plans were announced for a £200m pipeline network linking the Cormorant, Brent, Thistle, Dunlin and Hutton fields with Sullom Voe. The importance of this project can be gauged from the fact that the pipeline will carry up to one million barrels per day – one-third of the total oil output from the United Kingdom sector of the North Sea by 1980.

The events behind the choice of Sullom Voe as a multi-user terminal merit detailed attention. They date back to 1972, for in that year the county council, faced with an increasing number of oil-related planning applications and with evidence of the extent of oil and gas reserves in the East Shetland basin, commissioned two studies which identified Sullom Voe as the most suitable site within the islands for the location of a major oil terminal and/or transhipment facilities.

In November 1972 the county council promoted a provisional order which was intended to authorise it to acquire land (through compulsory purchase powers, where necessary) and undertake certain works, to operate certain harbour authority powers and to take equity capital in ventures established in the areas covered by the order. These extended to about 9500 acres in the Sullom Voe and Baltasound areas. Underlying this legislation was the belief that Shetland could do without North Sea oil but that, if the national interest insisted that oil and gas developments should come to Shetland, then they would be planned in such a way as to minimise disruption to the Shetland community and economy, and this implied that they would be persuaded to come together in one integrated complex at Sullom

Voe. The county council felt that the existing planning acts gave it inadequate powers to pursue such a policy and hence took the unique step of promoting a provisional order in Parliament to acquire the desired additional powers. With some modifications this became law (The Zetland County Council Bill, 1974) in April 1974.

Apart from its particular application within Shetland, the Act raises some fundamental issues regarding control of North Sea developments and may well set a precedent for similar action by other local authorities. Indeed, Orkney county council has virtually completed the same legislative steeplechase. The influence of central government, both Edinburgh and Whitehall, has been minimal and, whereas the local authorities in the North East and the Moray Firth have generally adopted a passive approach, Shetland has stepped into the void and armed itself with powers that no other local authority has. Indeed, this was reinforced recently by an agreement with various companies that closely resembles a barrelage tax on oil flowing through Sullom Voe and will provide the county council with an annual income of around £2m reducing significantly their dependence on central government finance. This again raises a precedent, the implications of which will not become clear in the immediate future, but which may be followed by other local authorities.

Shetland county council now has considerable powers over oil and gas developments through the operation of normal planning policies and the additional powers available under the 1974 Act. Developments are still at an early stage but are of two main types: supply bases; and landfall terminal facilities and related processing activities. The supply bases have been discussed above, and only general indications of the likely scale of development at Sullom Voe have appeared. It seems reasonable to expect, however, that oil and associated gas from the Brent group of fields, Ninian and probably Statfjord will all be piped to the Shetlands and this will necessitate substantial terminal and transhipment facilities, employing up to 600 men. Many of these would be concerned with maritime activities and the posts could be filled by Shetlanders without many problems. Gas-processing activities, such as a liquefied petroleum gas plant, are also likely but the prospects for similar oil-processing activities are not so obvious. The county council is planning for direct employment creation at Sullom Voe of between 550 and 1000, and although this may well seem small, it is immense in comparison with the size of the present labour force in Shetland.

Already signs of serious problems in the service sector are emerging and construction activity has only begun on a small scale. Shetland will have to depend largely on immigrant labour (including return emigrants) for the construction and permanent operation of the Sullom Voe complex, and this will create the same housing pressures that have built up in the North East and the Moray Firth. There is the added complication that the fish processing and knitwear industries are doing badly at the present time and, if the oil developments proceed rapidly, labour losses in these indigenous industries may be not only harmful but completely destructive. The crucial period will be during construction work at Sullom Voe. Furthermore, the complex may well exacerbate the problems of other areas in Shetland such as Northmavine, which lies north of Sullom Voe, and this is a danger that has received little attention to date.

Undoubtedly the county council has demonstrated a farsighted and vigorous attitude to the marriage of oil interests with those of Shetland. It appears to have provided a model framework for local authorities to use in such situations and its experience in the next ten years will be much more instructive than those of the North East and the Moray Firth councils. Whether or not its intentions are realised will become apparent during that period.

In Orkney the situation is similar but on a much smaller scale. One small supply base is in operation and a terminal is under construction on the island of Flotta to take oil from the Piper field. It is likely that oil from at least one other field (Claymore) will be piped to Flotta but this would not substantially increase the size of the terminal. Hence the problem of scale, which is the big danger in Shetland and East Ross, should not arise in Orkney, and the oil developments should provide a welcome and long-term diversification of the economy.

Rest of Scotland

There are three other large-scale developments under way elsewhere in the Highlands. On the Island of Lewis, at Stornoway, the shipping group Fred Olsen are constructing an engineering facility which could eventually employ up to 1000 people and will be very welcome in Lewis where the male unemployment rate has frequently exceeded 30 per cent in recent years. Across the water on the mainland, the Howard–Doris consortium has begun construction of a concrete production platform at Loch Kishorn in Wester Ross. Planning permission for this site was granted in the aftermath of the Secretary of State's decision to refuse permission to the Mowlem–Taylor Woodrow consortium for a site at Drumbuie on Loch Carron. This decision was taken mainly on social and environmental grounds, and, in view of the problems that large construction·labour forces create in remote areas, the Loch Kishorn site has been restricted to a maximum of 400 men. The Drumbuie case, which has attracted a great deal of attention, and the general strategy regarding the location of platform sites are discussed in chapter 7.

Further south in Argyll McAlpine are building three concrete platforms, but although Argyll is part of the Highlands much of it (including the McAlpine site) lies within the daily travel-to-work area of the Glasgow conurbation and consequently many of the problems arising from the need to import labour should be avoided. Indeed, government pressure is switching attention away from the Highlands to the east shore of the Clyde and various government-commissioned studies have identified possible sites there for platform construction. In that the platform builders are the largest single employers of labour this is very important for the Glasgow region and the Central Belt in general, but of more importance in the long run is the extent to which other aspects of the manufacturing phase build up in the region. As can be seen from table 6.1, to date this has happened only on a small scale, although any figures are likely to underestimate the true situation because of the difficulties in tracking down the large number of subcontractors to the main developments. Manufacturing activity in the Glasgow region is concerned mainly with the fabrication of modules and ancillary equipment (pumps, generators, cranes), and these activities should grow both spontaneously and in response to government exhortations, although the scale of employment creation is likely to remain small relative to the region's needs.

However, two shipyards on the Clyde, Marathon Manufacturing and Scott Lithgow, have lengthy order books for jack-up rigs and drill-ships. These are not intended for use in the North Sea but they demonstrate the export opportunities available in other off-shore areas, and it could well be that the Glasgow region will benefit more from such activity than from the North Sea. Certainly, involvement in the exploration and production phases is limited by locational characteristics but, given the range of engineering activity in the region, there is considerable scope for much greater involvement in the manufacturing phase, as was suggested in chapter 4. In addition, the headquarters of the Offshore Supplies Office has been located in Glasgow and it is likely that the proposed British National Oil Corporation will also be located there.

To date, more activity has taken place in the eastern part of the Central Belt, in the Edinburgh and Tayside regions, which embrace the Firth of Forth and the Firth of Tay. RDL North Sea have completed a steel platform for the Auk field and are currently building two other platforms. Two companies fabricating modules employ 500 people each and a pipe-coating yard, over 300. In addition, the terminal facilities at Grangemouth and Hound Point for BP's Forties field will come into operation soon. It is noticable, though, that the rate of oil-related growth in the Edinburgh and Tayside regions has slowed down, and it is unlikely that there will be any substantial expansions in the exploration and production phases. Future prospects are again related mainly to manufacturing activity (an example being a large tank fabrication facility under construction at Arbroath), but there are increasing signs that this part of Scotland is suffering from the North versus West competition for oil employment.

6.4 THE FUTURE LEVEL AND DISTRIBUTION OF EMPLOYMENT

So far in this chapter we have been concerned largely with historical developments and the present situation. It is valuable to take a look into the future and see if the patterns are likely to change and, if so, why. This is particularly important for those authorities concerned with planning in the North of Scotland where the problems of scale and speed of development are most serious.

Reverting to the fourfold classification of activity used earlier in this chapter – exploration, manufacturing, production, construction – it is possible to identify a fairly general temporal relationship between the various phases. By definition exploration activity has to precede the other phases; following the discovery of a commercially exploitable oil or gas field, the manufacturing and temporary construction phases will last for three to seven years; and finally the production phase will last for a period of up to thirty years. Inevitably the phases will overlap as new discoveries are made, but the overall picture is already becoming much clearer. It is obvious that the future prospects of different areas are related to different phases, and, as was pointed out in the introductory section of this chapter, this means significant changes over time in employment levels and the nature of employment creation.

Taking the exploration phase first, on present evidence it appears that we are approaching the period of peak activity in the near future. Since the four rounds of licence allocations the exploration companies have been drilling their prime prospects and a common view in the industry is that the best prospects will have been drilled by late 1975 or early 1976. If this is the case, then success rates will begin to decline as companies turn to the less promising geological structures and a likely consequence is that the general level of exploration activity will fall. Forecasting future levels of employment creation during the exploration phase has to be based on forecasts of the number of rigs operating (i.e. using the ratio of 260 jobs per rig), and this is a precarious exercise, dependant as it is on future discoveries and on whether or not the government licenses new areas. On present information however the best estimates that we can make of the peak number of rigs operating in the UK sector each year are: 1974, 30; 1975, 40; 1976, 50; 1977, 45; 1978, 42; 1979, 40 and 1980, 32. With the introduction of a new fleet of semi-submersibles there will not be the large seasonal fluctuations that occurred in the early 1970s so there are unlikely to be serious discrepancies between peak and average figures. These estimates produce the employment figures for the exploration phase given in table 6.3. It appears that employment in this phase is likely to rise from its present level of 6000 to a peak of around 9500 in 1976 with a steady decline thereafter.

TABLE 6.3 ESTIMATES OF NORTH SEA EMPLOYMENT CREATION IN SCOTLAND

	Exploration	Manufacturing	Production	Construction	Total
1974	6000	8700	800	3500	19,000
1975	7700	9600	1000	4800	23,100
1976	9500	10,000	1500	6000	27,000
1977	8600	9000	2100	6000	25,700
1978	8000	8500	3000	5200	24,700
1979	7300	8000	3800	4000	23,100
1980	5800	8000	4500	3100	21,400

The manufacturing phase follows behind the exploration phase, lasting on average for five years after a successful discovery, but in fact the peak period of employment will not be five years after the peak exploration year for two main reasons. Firstly, as was discussed in chapter 4, a large element of activity is taking place outside Scotland – in north-east England, for example, but, more importantly, in the United States, Norway, France etc. – and these other areas appear to be providing the capacity to cope with fluctuations over time. Manufacturing activity in Scotland, therefore, should be more stable than would be expected – albeit at a lower level. Secondly, the increasing preference for concrete rather than steel platforms has resulted in a fall in labour requirements – by up to 60 per cent for individual platforms – and as platform construction is the largest single activity this means that the peak level will not be much higher than the current level and that employment will remain fairly steady throughout the decade. In table 6.3 it is estimated that employment in this phase will rise from its 1974 level of 8700 to approximately 10,000 in 1976 but will still be about 8000 in 1980. Furthermore, if the share of Scottish and UK involvement in North Sea markets can be increased employment requirements would rise correspondingly.

The construction phase runs in parallel with the other phases and will be susceptible to the same cyclical trends. A pattern of concentration in the Shetlands, Orkneys and the North East is becoming increasingly firm and the Sullom Voe proposals are a good example of economies of scale. As activity in this phase is related largely to the development plans of known discoveries, for the period up to 1980 it is fairly easy to forecast future employment levels, which suggest that again the peak will probably be reached in 1976 or 1977. These forecasts disguise to some extent the fluctuations that are inevitable with pipe-laying, terminal construction etc.

The final phase, the production phase, is by far the most capital-intensive, and employment levels will fall when this phase begins to supersede the other phases. Also, as was discussed earlier, virtually all North Sea oil will be refined outside Scotland so the potential level of employment during this phase will be considerably lower than might be anticipated. As production activity will build up slowly during the 1970s, the real effect of this decline in employment may be disguised but eventually it will be substantial. For example, when the Forties field is fully operational in the late 1970s, it will need a permanent labour force of approximately 700, compared with a peak of 7500 during the manufacturing and construction phases. Similarly, the Piper field will create about 300 permanent jobs, compared with a peak of 4000. The implications of these substantial changes over time should not be underrated. On the basis of known and expected development plans for both oil and gas fields, it is estimated in table 6.3 that employment during the production phase will rise steadily from the 1974 level of 800 to about 4500 by 1980 and will continue to increase throughout the 1980s. Once again, these estimates are subject to the usual reservations and should be treated with caution but the overall picture which they present is a justified one.

The relatively low level of labour demand during the production phase has important implications for the overall level, as can be seen from table 6.3. The table suggests that peak employment creation will probably be in 1976, in line with the peaks in exploration and manufacturing, and that employment will fall slowly but steadily thereafter. It may well be that in the event the peaks of individual phases will be delayed by a year or so, but there can be little doubt that once the production phase emerges as the dominant activity then direct employment will fall. To the extent that indirect, service, jobs are related to the original direct employment creation, the magnitude of the decline will be greater.

This has implications not only for the Scottish economy as a whole but also for individual areas. Within Scotland, areas such as East Ross and Fife are virtually wholly concerned with the manufacturing phase and there is a strong possibility that in such areas oil activity will have a limited lifespan of around ten years. East Ross has already experienced substantial fluctuations in activity and the consequent social and economic problems. Other areas will be affected similarly. In table 6.1 it was shown that a sizeable proportion of oil-related manufacturing activity is located in the Central Belt of Scotland and this proportion should rise in response to government pressure, but for this part of Scotland the North Sea offers limited opportunities and other off-shore markets must be found if employment levels are to be maintained in the 1980s. Over time, therefore, it appears probable that the present domination of the North of Scotland will continue and that its share of direct employment creation will remain between 70 and 80 per cent of the total.

6.5 CONCLUSIONS

In terms of direct employment creation arising from North Sea operations, we have estimated that by the late 1970s about 27,000 jobs will have been created in Scotland. In addition there are substantial opportunities in other export markets – e.g. for the construction of exploration rigs – and involvement in these markets could increase the number of direct jobs to between 30,000 and 35,000 if all off-shore activity were included. This employment creation in itself will generate additional indirect employment through the operation of the regional multiplier. For small, rural areas like the Shetland and Orkney Islands the value of the multiplier will be low, but for the whole of Scotland the increase in direct employment could eventually generate another 30,000 to 35,000 jobs in the service sector, bringing the total (direct and indirect) increase in employment to between 60,000 and 70,000 jobs in Scotland.

One aspect of major importance is that much of the employment creation is taking place in the North East and the Highlands and Islands, which are remote from the industrial heart of the country. In fact large developments are occurring in small communities where the present infrastructure and labour reserves are very inadequate. Consequently, serious economic and social problems have arisen in these areas as a result of the scale and rate of build-up of particular developments. It will be possible to divert some manufacturing activity – e.g. the construction of production platforms – away from these areas to the Central Belt, but it is unlikely that the North East's and the Highland's share of oil-related employment will fall below 70 per cent. Their proximity to the discoveries and the locational requirements of the exploration and production phases leave little room for manoeuvre. It is, therefore, essential both to the long-term future of the North of Scotland and to the maintenance of North Sea programmes that solutions are found to the problems that have arisen.

Over time, it is likely that employment will fall slowly but steadily after the peak has been reached. This is largely because the labour requirements of the production phase are less than those of the other phases and because very little North Sea oil will be refined in Scotland. In addition, it must be recognised that during the various phases and in particular areas there will inevitably be substantial fluctuations in the level of activity – the best example being that of platform construction in areas like East Ross and Argyll. The changing technology of the North Sea therefore creates challenges both off-shore and on-shore, and the latter poses additional difficulties for economic planning and regional policy which are considered in the following chapter.

NOTES TO CHAPTER 6

1. Quoted in E. Richards *The Leviathan of Wealth* London, Routledge (1973) p. 173.
2. Practices vary from rig to rig but most operate one week on–one week off, two weeks on–one week off or two weeks on–two weeks off systems.
3. *Education and training for offshore development* London, HMSO for Department of Employment (1973).
4. See, for example, *Scottish Economic Bulletin* no. 6 (July 1974) pp. 12–14.
5. See table 5.3, p. 96.
6. In this context it is interesting to note the postponement of BP's expansion plans for their Grangemouth refinery.
7. In Aberdeen city itself the 1974 rate of unemployment was below that for south-east England. This is the first time that *any* employment exchange in Scotland has had an unemployment rate below the south-east England level since detailed unemployment statistics became available in 1923.
8. H. W. Richardson *Regional growth theory* London, Macmillan (1973) p. 29.
9. *North East Scotland and the Offshore Oil Industry* Aberdeen, NESDA (1974).
10. *Housing in the Aberdeen Area* Aberdeen, NESJPAC (1972).
11. J. Francis and N. Swan *Scotland's pipedream* Edinburgh, St. Andrew Press (1974).
12. J. Francis and N. Swan *Scotland in turmoil* Edinburgh, St. Andrew Press (1973).
13. Highlands and Islands Development Board *First Report* (1967) p. 4.

CHAPTER 7

Government Policy
on Land

*A little neglect may breed mischief ... for want of a nail,
the shoe was lost; for want of a shoe the horse was lost; and
for what of a horse the rider was lost.* (Benjamin Franklin,
Poor Richard's Almanac)

7.1 INTRODUCTION

It may appear somewhat arbituary to distinguish between the on-shore
and off-shore aspects of government policy, but there are markedly
different implications which are better considered separately. Many
of the on-shore effects discussed in earlier chapters are directly
related, of course, to factors such as the rate of production off-shore
and government licensing policy. Thus it would be impossible or
ill-advised to pursue very different policies in these two spheres.
Consistency of approach is important although the process of policy-
making does not lend itself easily to the achievement of this aim.

For the purposes of this chapter we have identified three categories
of policy on-shore:

1. policies specifically directed to the North Sea developments
(licensing policy, control over the rate of production etc.);
2. policies wholly concerned with Scotland (and operated by
Scottish bodies, such as the Scottish Development Department and
the local authorities, particularly in matters of physical planning);
3. national economic policies which operate regardless of the North
Sea developments (regional policy, incomes policy etc.).

Any evaluation of these policies or suggestions as to their future
application has to take into account three major factors. First, the

138

scale, pace and sheer unpredictability of North Sea developments in the past have inevitably created major difficulties on land. Some of the mistakes made can now be clearly identified, but it is always easier to construct a blueprint for the past than for the future. The future is still bedevilled by extreme uncertainty, and policy has to be constructed in this light. Its watchword should be flexibility. Secondly, the problem of adjusting to North Sea developments has been exacerbated by the fact, emphasised in the preceding chapter, that they have largely fallen on communities in the North of Scotland which lack the urban and industrial base easily to accommodate rapid growth. Geography makes this problem unavoidable and, indeed, we shall argue the need to face and accept it more readily. Thirdly, one inevitable consequence of North Sea developments has been the creation of conflicts of interest. The imperatives of the UK balance of payments have dictated a policy at sea which has emphasised the need for rapid exploration and production and its very success, at least in the exploration phase, has placed major strains on the social and economic fabric of the communities most affected. Planning and foresight cannot make these conflicts disappear entirely, but they may allow some compromise between different interests.

Our major task in this chapter is not to pick over past decisions so as to underline the many difficulties that have arisen. For the reasons we have stated, we consider that many of the problems were unavoidable and simply reflect the impossibility of correctly anticipating the size of North Sea finds. Rather we are mainly concerned, given these finds and possible future discoveries, to map out the broad outlines of the development strategy which might be appropriate for Scotland in the future. In the following section we consider licensing and production policies and their possible impacts on land. This is followed by a consideration of the merits of 'dispersing' and 'concentrating' North Sea activities and then by a discussion of one set of decisions which has major economic and physical planning implications – the siting of production platforms. The analysis is completed by sections discussing the financing of infrastructure developments and two aspects of UK economic policy – incomes policy and regional policy – before the concluding arguments are summarised in the final section.

7.2   LICENSING AND PRODUCTION POLICIES

A crucial determinant of the scale and lifespan of on-shore impacts is obviously the rate of exploration and development in the North Sea. This determines the demand for supply bases, production platforms, terminals and the like, and consequently variations in the level of off-shore activity will result in varying impacts on-shore. These on-shore impacts, however, are not (or should not be) the main factor in determining the level of off-shore activity although they should be taken into consideration.

There are three aspects to this matter:

1. the extent to which the level of off-shore activity can be regulated by government policy;
2. the precise relationship(s) between off-shore activity and on-shore impacts;
3. the feedback effects that delays, bottlenecks etc. on-shore have on off-shore activity.

Firstly, the policies discussed in chapter 2 set the framework within which North Sea activity takes place. Of the existing armoury of government powers the issuing of exploration and production licences is the most influential. The government is committed to extending its powers to control the level of production in the national interest, although it has been stated that this will not affect its determination to build up production as quickly as possible over the next few years. Control is much easier in the medium term than in the short term, and it is difficult to see how the government could radically alter the pace of developments under way already. It seems likely, therefore, that when the desired level of production has been reached – presumably in the early 1980s – measures will be taken to maintain fairly stable production levels from then on. From the on-shore point of view stability is attractive, but it must not be forgotten that physical events such as actual discoveries and the production characteristics of fields inevitably make it very difficult rigidly to apply particular policies. Similarly, UK objectives may well be frustrated by events elsewhere in the world which make other off-shore areas more attractive and divert resources away from the North Sea. This is a threat which crops up with increasing regularity and which cannot be ignored, as to a considerable extent the North Sea should be used as a proving ground for British equipment and technology which can then be used in other off-shore areas. It is our view that there is no case for introducing any protectionist policies for the UK sector of the North Sea and, if this view is accepted, domestic policies must be more susceptible to external pressures – even on the desired level of production. Where government policy may be more influential domestically is in the control of production activities, processing and refining, by insisting, for example, that a certain proportion of North Sea oil is refined in the United Kingdom. This would be similar to the controls exercised over natural gas production and in specific geographical areas this would have important effects.

Secondly, it is essential to remember that it is impossible to identify a precise relationship between what occurs off-shore and consequently on-shore. As was shown in chapter 3, conditions vary so much from field to field and from company to company that it is impossible in advance to detail on-shore requirements. In particular, the consequences of discovering hydrocarbons off the remote areas of Scotland are very different from those that would have occurred if the discoveries had been made close to large centres of population and economic activity. The ability to react flexibly in changing circumstances must therefore be implicit in any government policy in this sphere.

Thirdly, the process is not one-way: many on-shore activities have reciprocal effects at sea. The most noticeable effects are those arising from delays in fabricating production platforms. Virtually all the development programmes for the oil and gas fields have suffered delays of this nature and the main causes have been the great pressures put on local labour markets and infrastructure, chiefly housing. If future programmes are to proceed on schedule, it may be necessary to ensure first of all that conditions on-shore – e.g. in the areas where platforms are being built – are adequate. There are signs already which suggest that in the short run Scotland could not satisfactorily sustain a rate of development much higher than is being achieved at present.

It has been argued that a slowing down of the rate of development would enable British industry to increase its involvement. As was discussed in chapter 4, there is little evidence at the moment to support this view, and our own belief is that the disappointing performance of domestic industry is attributable to other factors. A stronger case can be made on the grounds of the strains being placed on the productive capacity of the North of Scotland; and, although the solution of many of the problems discussed in chapter 6 is in hand, there are lessons to be learned for other areas if discoveries are made elsewhere in UK waters.

7.3   DEVELOPMENT STRATEGY WITHIN SCOTLAND

Most of Scotland's economic problems are concentrated in the Central Belt and particularly in the Glasgow region; in contrast, most of the North Sea employment creation is taking place in the North and the East. Almost inevitably, the Scottish Office, supported by central government, have endeavoured to bring the two together by persuading as much oil-related activity as possible to locate in the Central Belt. Essentially there are two ways of doing this: through the provision of inducements or incentives for firms locating in preferred areas; and through the positive discouragement of other locations. The powers for pursuing the former objective are economic; those for the latter, physical planning.

This issue of concentration or dispersal is crucial and it is worth reiterating the arguments for each case. Scottish Office policy is summarised in a coastal strategy document issued in 1974 by the Scottish Development Department.[1] This sets out various guidelines for local authorities and prospective developers regarding the location of oil-related activities in Scotland and distinguishes between preferred development zones and preferred conservation zones. The underlying principle is that it would be beneficial to group developments together in or near existing population centres as this would have various advantages:

1. the avoidance of a scatter of industrial developments, affecting many small communities and numerous rural areas;

2. fuller use of existing labour pools, housing and public services;

3. economic provision of additional services needed for the new developments;

4. greater likelihood of diversification to cushion any subsequent decline;

5. the opportunity to use new developments to rehabilitate obsolescent areas;

6. greater likelihood of existing businesses being able to adapt to serve new needs.

On this basis various development zones were identified as having some or all of a list of desired features:

1. a community or series of settlements which can be expanded without incurring a future danger of severe economic or social decline, resulting from over-dependence on one source of employment;

2. some flat land on the coast and in the hinterland which could absorb major development;

3. suitable ports and harbours with some potential for developing the dockside land;

4. existing communications and infrastructure or the potential for improving them economically;

5. areas in which economic and environmental rejuvenation is required;

6. areas in which operations associated with oil and gas could be grouped.

Briefly, the development zones chosen gave priority to the Clyde, Forth and Tay estuaries and included those areas in the North East and the Highlands and Islands where developments were taking place already. The whole of the west coast of the Highlands was not zoned for development.

Two simple points can be made immediately. First, the basic strategy is undoubtedly sensible and it is unfortunate, although understandable, that it could not have been applied a few years ago, for quite a few of the existing development zones would have had difficulty in getting on the list of preferred zones. Secondly, the application of the strategy is much more important than its theoretical underpinning and the actual application so far is open to quite a few objections. Our objections to the policy of diverting the oil industry from its natural predilection for the North of Scotland to locations in the Central Belt are both general and specific. We shall deal with the general, and more theoretical, arguments first and then look at specific examples.

In the context of the oil and gas developments in Scotland, the overriding objective must be to build up a vigorous, profitable and internationally competitive off-shore industry. What Scotland has lacked for many years has been growth industries, although the complex of computer-based firms in the Central Belt (mainly US-owned) provided this to a limited extent in the 1950s and 1960s. Scotland has long been a proponent of the policy of growth areas which implies that, *inter alia*, the establishment of growth centres or growth areas will increase regional growth rates, and it is a policy that could well be applicable to the off-shore oil industry in Scotland.

The definition of a growth centre and the conditions that must exist for such a centre to emerge or be established is often vague or misunderstood in planning documents. Growth centres are much more economic phenomena than geographical and their essential feature is that of an industrial complex existing in a fairly narrowly defined geographical area. Such a complex must consist of a group of firms which have important economic or technical linkages and which are preferably of a large size and part of a fast-growing industry or industries. Without the existence of these inter-industry linkages there is little reason to expect one geographically concentrated group of firms to have a higher rate of growth than any other similarly concentrated group. A growth centre must generate significant internal and external economies. An EEC study of the growth centre in southern Italy made the following crucial point: [2]

> In order to produce, a manufacturer must have close at hand all the intermediary industries: sub contractors, i.e. firms concerned at one stage or another in the manufacture of a specific article, and suppliers of services, in particular those whose function is installation and maintenance of plant. None of this intermediary work will be a paying proposition if it is done for a single customer. Each intermediary manufacturer must devote himself to one operation only, but he must do this for a large number of user industries. Only then can he obtain the production level necessary to bring his costs down low enough to justify his existence. Consequently, it is quite clear that for complex-cycle industries an entrepreneur can only reasonably envisage manufacturing a finished product in centres where he can find all the industries auxiliary to his own. Conversely, a subcontractor or supplier of services will only set up in an area with an adequate market in the shape of firms requiring his specialized services.

The creation of such external economies may well be a prerequisite of the establishment of a strong and competitive off-shore oil industry in Scotland. As chapter 6 showed, there are signs that this is happening in the Aberdeen area with the appropriate mix of linked firms coming together to form what might be loosely described as a growth centre. This would increase the magnitude of the indirect and induced effects arising from the original increases in income and employment. It is our view that, within the existing administrative set-up in Scotland, these indirect and induced effects may be as important as the direct effects. It seems that, left to its own devices, the off-shore oil industry will increasingly cluster together on the east coast of Scotland, particularly in the Aberdeen area. If, as is the case, the Scottish Office feels that it is desirable to decentralise activity, then it is necessary to identify very carefully those activities that can be located elsewhere without destroying the generation of external economies which appears crucial to the long-term establishment of the off-shore oil industry. There is then a potential conflict between long-run economic efficiency and more short-run social considerations of alleviating high unemployment.

In that the policy of decentralisation has had little effect to date, our fears may turn out to be unfounded, but recent government decisions to locate the headquarters of the proposed British National Oil Corporation in Glasgow and the new school of drilling technology elsewhere in the Central Belt suggest that inadequate consideration is being given to the points outlined above. Both these institutions are prime examples of activities that would be expected to add to external economies in the Aberdeen area; in terms of the oil industry in the Central Belt there is very little to which the benefits of these institutions can be added.

It may appear that our desire to follow a type of growth centre policy in respect of the North Sea developments (i.e. by building on the industry's expressed preferences for locations in the North of Scotland) sits uneasily with our previous emphasis on the pressures that have been imposed on the areas affected, and that what we are suggesting in this chapter would add to the problems described in chapter 6. To some extent this is true, particularly in the short run, but it is our view that solutions can be found to these problems and that in the long run benefits to Scotland through the establishment of a vigorous and viable off-shore industry will outweigh the costs currently being incurred.

7.4 LOCATION OF PLATFORM SITES

It is instructive to look in detail at the issues surrounding the construction of production platforms, partly because of their crucial importance to development programmes and partly because of the considerable difficulties that they have created for the planning process in Scotland. Many of these issues were embodied in the inquiry into the Mowlem–Taylor Woodrow application to build concrete platforms of the Condeep design at Drumbuie on Loch Carron, one of the lochs on the west coast.

Nothing has happened in fact at Drumbuie but it provides the best example of the range of issues surrounding the impact of large oil developments on the small communities of the Highlands and Islands. It has become the *cause célèbre* of North Sea developments, as it pitted a small community of twenty-three people, the National Trust for Scotland and the 'environmental lobby' against two large companies, the Department of Trade and Industry (and then Energy) and the Highlands and Islands Development Board. Both sides could muster formidable arguments.

In April 1973 two companies, John Mowlem and Taylor Woodrow, separately applied for planning permission to build concrete production platforms at Port Cam, Drumbuie, on land owned by the National Trust for Scotland. Subsequently the companies came together in a joint venture to construct platforms of the Condeep design which have proved a frequent choice of the oil companies and are being constructed in Norway.

The issues surrounding the planning application were on the surface straightforward. On the one hand there was a demonstrated need for concrete platforms; the Condeep design was popular with the oil companies; and the deep-water construction requirements made Drumbuie the most suitable site in the United Kingdom. On the other hand Drumbuie was a small west coast crofting community, and a development of the scale and nature proposed would produce irrevocable changes in an area of outstanding natural beauty. A complicating issue was the fact that the land had been left inalienably to the National Trust for Scotland and it would have required an Act of Parliament to allow it to be sold or leased to the construction companies.

In the light of the large number of objections to the proposal (including one from the National Trust), the Secretary of State for Scotland ordered a public inquiry to be held and this lasted from November 1973 to May 1974. The inquiry was widely regarded as a test case. In the event, in August 1974 the Secretary of State ruled against the proposal, in line with the recommendations of the

Reporter who conducted the inquiry. Unfortunately, the reasons for refusal are obscure and the decision offers little guidance to local authorities faced with similar applications. Hence it is necessary to look at the decision and related arguments in more detail.

Apart from the two companies involved, the main proponents of the Drumbuie project were the Department of Trade and Industry (as it was) and the Highlands and Islands Development Board. The Department of Trade and Industry (DTI) presented a detailed memorandum at the inquiry, which purported to show that there would be a need for up to 50 production platforms to be installed in the UK sector of the North Sea by 1980 and that about half of these structures would be of the concrete gravity type. The Drumbuie site could produce two structures per year – i.e. 20 over a ten-year period – and at £20m per platform these would result in a total £400m of orders, with 80–95 per cent of the materials, labour and equipment coming from UK sources. If concrete platform orders were lost to Norway and elsewhere this would be of considerable detriment to the United Kingdom economy and balance of payments.

The DTI's evidence can be criticised on three main counts. The first is the overriding necessity of the Drumbuie site because, although it is probably the best site of its type, the Mowlem–Taylor Woodrow consortium has since applied for a site in Argyll. The second main criticism concerns the estimates of platform demand, which many people believe to be too high and hence to have artificially strengthened the claim for a site at Drumbuie. Lower estimates have been produced by the present authors[3] but in fact the DTI has refuted its own estimates (perhaps unknowingly). In a statement on platform demand and platform sites issued in August 1974 the newly formed Department of Energy said that on average a production platform was expected to produce about 5 million tons of oil per year. Taking into account gas and pumping platforms (reckoned to number between 5 and 10 by 1980), the DTI's evidence at Drumbuie implies a 1980 production rate of 200 million tons of oil, compared with the official estimates of 70–100 million tons at that time! Even the August 1974 estimates are contradictory, implying a production rate of between 250 and 300 million tons by 1980, compared with the revised official estimates of between 100 and 150 million tons and stated government intentions to slow down the rate of production. The third main criticism is also related to the platform demand estimates, in that the DTI forecast a peak annual demand for 6 concrete production platforms, and maintained that there was a need for capacity to meet that peak demand, which in turn implies that in the non-peak years quite a few sites would be without orders and there would be considerable redundancies. As was discussed above in the context of East Ross, such fluctuations in employment are disastrous in the Highlands (though not necessarily so in the Glasgow area or the industrial areas of England) and particularly so in small and isolated communities such as Drumbuie.

In the event the Reporter for the inquiry recommended refusal of the proposal and he was supported by the Secretary of State. A rush of new applications in and around the Firth of Clyde area followed and the Scottish Office and the Department of Energy both commissioned consultants to identify suitable sites. The consultants' report to the Department of Energy selected four sites: the first conflicted with a planning direction given previously by the Secretary of State; the second embraced a bird sanctuary; the third was earmarked already for an iron ore terminal; and the fourth does not appear to have a wide enough channel for floating out platforms.

In the meantime the government has announced its intention of taking platform sites into public ownership and leasing these to companies with approved designs, of which seven were listed in the original August 1974 proposal. It is not clear how this will affect sites in use at the present time, particularly the steel sites on the east coast, or what the position will be regarding companies who obtain orders for non-approved designs. Normal planning procedures are continuing, the Secretary of State has called in certain applications for his own consideration and more public inquiries are likely.

All this may be very confusing to follow and it has been presented as such to show the dangers in not defining and pursuing a firm and clear policy. A concentration of concrete platform sites in the Clyde area would be very desirable on balance, providing welcome employment and reducing the pressures on remote communities in the Highlands, but the policy must be clear for all to see and must take account of both the state of technology and existing capacity. It is vital to base the policy on the most accurate estimates of platform demand possible and the Department of Energy's estimates are open to far-reaching objections. Similarly, the possibility of a more widespread adoption of sub-sea completions or tension-leg platforms has to be considered fully.

In this context there is a strong case for simplifying and speeding up planning procedures; and this need not be done at the expense of the proper consideration of the range of issues involved. As we implied above, the public inquiry system has been of little help in devising an appropriate policy for the location of platform sites. We feel that it could be replaced by simpler procedures operating within guidelines laid down by the Scottish Office and central government. The coastal planning strategy is a case in point. We also believe, however, that much more executive power should be given to local planning authorities – despite the discouraging experience in East Ross – and that, once they are provided with reasonable advice on the probable outcomes of alternative policies, they should be free to decide in accordance with their interpretation of local interests. In doing so they should be able to insist on detailed information being produced by prospective developers which should reduce the number of speculative projects. Only when there is an evident and overriding case for the imposition of the national interest should central government intervene.

7.5   PROVISION OF INFRASTRUCTURE AND FINANCE

As we have seen, North Sea activities have been highly concentrated geographically and the scale of the impacts has often been large relative to the size of the communities affected. The future geographical pattern of employment is likely to be much the same and, indeed, we have argued the case for encouraging this process in certain directions. The desirability and the practicability of such a policy turns on whether these communities, and particularly the Aberdeen area, can adapt to the demands that might be placed upon them.

Most of the problems encountered in the Aberdeen and East Ross areas have sprung from a lack of spare infrastructure capacity (roads, harbours, airports, schools, houses) and from the difficulty of providing additional capacity quickly. Such investment in the infrastructure is necessary to allow for the immigration necessary to meet the additional demand for labour, and if it is not provided then established industries will inevitably experience substantial difficulty in adjusting to an increasingly competitive labour market situation.

The lack of infrastructure provision in the past is a consequence of many factors, among which real supply side restraints have been prominent – labour shortages, particularly of skilled construction trades; a lack of zoned and serviced land; etc. However, as is always true of the construction industry, these restraints do become less pressing with the passage of time and there is already clear evidence that the capacity of the industry has now significantly increased. This is true even for housing, which has presented the major difficulties in the immediate past. Yet other problems may remain – in particular the question of how the costs of new infrastructure investment should be shared between private companies, local communities and the central government. This is the issue that chiefly concerns us in this section.

Taking the first point, there is a strong case for re-examining the principles that determine the distribution of costs between local authorities, central government and private developers. At the present time much of the burden of providing new houses, schools, roads, etc. is falling on relatively small local authorities who clearly are experiencing severe financial problems in endeavouring to meet the demands placed upon them. Borrowing is the main solution, but this is costly in a period of inflation and difficult because of uncertainties about future revenue from increased rateable values. The process of local government reorganisation in Scotland further complicates the problems in that responsibilities are being handed over to new authorities.

Many of the oil and gas developments are occuring in areas with little or no infrastructure or where the historical process of decline and emigration has produced a situation in which much of the public infrastructure requires renewal. Shetland and Orkney are the prime examples of this, as is Loch Kishorn and as would have been Drumbuie. To avoid discussing the merits and demerits of particular situations, it is useful to look at the issue in abstract terms, taking Drumbuie as the example. Suppose that planning consent is given for a production platform site at Drumbuie. Such a decision would involve certain environmental and social costs (disruption and even demise of small communities, the loss of scenic value) and certain economic costs (loss of tourist income, the need to provide infrastructure for a project of limited duration) which would certainly be higher than those that would be incurred by the use of other sites. This is particularly true for infrastructure costs because in remote communities there is generally little spare capacity in terms of houses, roads etc. and the new investment would be likely to be seriously underused when production platform construction came to an end. The same argument would have much less force at a site such as Hunterston where less investment in infrastructure would be necessary and where there would be a strong possibility that production platform construction would be followed by other types of industrial activity.

A private developer using the Drumbuie site would therefore impose costs on the community which are likely to be substantially greater than the external costs imposed by the use of an alternative site – and this seems to have been accepted by all the expert witnesses to the Drumbuie inquiry. On the other hand, it is also agreed that the Drumbuie site would have certain unique characteristics which make it particularly suitable for production platform construction. These benefits would accrue, of course, to the site user. We can then formulate the principle, which is admittedly easier to formulate than to apply in practice, that the site should not be used for production platform construction unless the economic rent acquired from the use of the site is sufficient to compensate for the external costs which the use of the site imposes on the community. Further, the gainer, in this case the private developer, should compensate the losers. One method of achieving this would be to estimate as far as possible the external costs imposed by the use of the site and then to auction the site subject to the condition that its sale should be sufficient to cover these extra costs.[4] Of course, such a system has very substantial imperfections in practice. How does one establish the losers? How are they to be compensated? What value should be placed on the 'way of life' of small communities? What is the price of a beautiful view? Any process of compensation must be very imperfect, but this is no excuse for ignoring the principle that as far as is possible the external costs of such developments should not be borne by the community but by the enterprises that make such developments necessary.

In the Drumbuie case, and in most cases where production platform activity is concerned, the developments that impose the unwanted externalities can be fairly identified and the duration of the developments might be relatively short. At the other extreme of the spectrum we could take the case of the city of Aberdeen where more than 200 new companies are involved and where the time scale of developments will be substantially longer. In this case it would seem that local authorities should bear the capital costs as they will benefit over the long run through increased rateable value. The capital expenditure required in the short run to make these developments possible can then be financed by borrowing.

In practice there may be considerable difficulties. Firstly, local authorities obtain rate support grants which are equalisation grants. As rateable values increase, rate support tends to diminish on virtually a one-for-one basis, transferring the costs of meeting higher capital expenditure from the central government to the occupiers of new premises. Secondly, to finance new construction by borrowing will, at present interest rates, bring an appreciable increase in the level of current expenditure and hence in the sums that it is necessary to raise through rates. Thirdly, given the existing methods through which controls over capital expenditure are exercised, an increase in borrowing consents for one area is at the expense of another. Fourthly, there is, except for the new towns, no method of looking at the needs of an area as a whole as the allocation of consents is by function (roads, houses, schools etc) and not by area. There is no doubt that the Scottish Office is acutely aware of the necessity for considering capital expenditure by area given the particular circumstances arising from North Sea oil developments, and it has already in practice made substantial efforts to adjust to the new capital expenditure needs of authorities in the Highlands and Islands and the North East of Scotland.[5] However, the whole process merits a good deal of further study. In this instance the powers that have been acquired by Zetland county council and the substantial income they will derive from oil developments provide an example of what can be done to balance more fairly the costs and benefits of particular developments.

7.6 APPLICATION OF NATIONAL POLICIES

Incomes policy

Policies for the North Sea, both off-shore and on-shore, cannot operate in a vacuum detached from economic conditions in the UK as a whole. As has been suggested above, there will be times when the UK interest weighs strongest and conflicting local or Scottish interests have to be overruled. Equally, there will be times when a modification of UK policies is desirable in the interest of Scotland or particular areas – and possibly, in the end, to the benefit of the

UK. It is impossible to lay down firm guidelines on this matter but we would press very strongly the case for the flexible interpretation of national economic policies when they impinge on North Sea developments.

A good example of this is the last experiment in incomes policy which was concluded in the summer of 1974 but which may well be tried again in the near future. If this were to be more successful and durable than the incomes policies of the past then lessons must be learned from previous mistakes – and the rigorous application of incomes policy in the North East and the Highlands and Islands was surely one of these.

In 1971, at the outset of the North Sea developments, both these regions were low-wage areas.[6] The pace and scale of expansion in these areas resulted in increased competition for labour and a rapid exhaustion of the labour reserves which had been reflected in relatively higher unemployment and low activity rates. The normal market response to this would be a rise in the price of the scarce resource, in this case labour, in order to attract, through immigration, in increased supply. In addition to requiring a change in the spatial distribution of labour, a period of rapid economic change usually requires reallocation between different industries and occupations. The most obvious example in this case was the need to attract more labour into the construction industry in order to provide the additional infrastructure which immigration required.

Unfortunately, the incomes policy applied in the UK from November 1972 to May 1974 allowed neither of these adjustments to take place. There were three phases: phase 1 lasted from November 1972 to March 1973 and imposed a stand-still on all pay increases and other improvements in the pecuniary terms and conditions of employment; phase 2 ran from March 1973 to November 1973 and permitted, with no exception, increases of £1 per week plus 4 per cent up to a ceiling of £250 per annum; and phase 3, which was terminated in May 1974, allowed increases of £2.25 per week or of 7 per cent whichever was greater, up to a maximum of £350. In addition, in phase 3 there were certain exceptions for 'flexibility', efficiency and unsocial hours.

None of these phases allowed for exceptional wage increases to reflect labour scarcity. In this, the last major experiment in incomes policy has to be sharply distinguished from its precursors, as each previous experiment had recognised the need for exceptional treatment in conditions of labour shortage. It was then impossible to raise the general level of wages in low-wage areas, or to raise relative wages in an industry or occupation where labour shortages were acute. The incomes policy therefore prevented the one form of adjustment which guarantees an effectively functioning labour market.

This in itself would inevitably have created difficulty in those areas experiencing the impact of rising oil-related employment. Yet there was one loophole in the operation of the policy which further aggravated the situation and threatened the position of established firms. While it was not possible for 'indigenous' or 'established' firms to raise wages relative to each other or to raise the general level of wages relative to other areas, it was possible for new firms entering the area (in the oil industry or in other sections) effectively to evade the incomes policy restrictions. New firms were required to 'pay the going rate', but as the going rate could not be defined such a clause proved impossible to operate. Many firms were employing labour in occupations new to these areas. Even if this was not so, it is well established that in any sizeable labour market there is a substantial wage range between the wages paid to labour for the same occupation.[7] The new firm could then set wages at whatever level was necessary to obtain the labour it required, and established companies could not respond. Not surprisingly, many established firms attempted to evade the policy. Some were caught, but there was no way of curbing the wages offered by incoming firms.[8]

The failure to make an allowance for a situation of rapid economic change therefore compounded inefficiency with inequity. It prevented a general rise in wages, but allowed particular firms considerable latitude. The distribution of labour that resulted was then capricious. Relaxation of the policy would not have provided any easy solution for established companies, who would have had to meet increased competition and hence higher wage costs, but such a rise is a necessary, if not a sufficient, condition to obtain additional labour through immigration. The other necessary condition is a rise in wages in the construction sector. It is perhaps fitting commentary on the absurdities of the incomes policy as applied that in the final phase the Pay Board was attempting to reduce wages in eight building firms, all of whom were already short of labour.

An unwillingness to contemplate exceptional treatment is understandable under a wage freeze period such as phase 1. However, it is a sad reflection on our understanding of labour economics that in phases 2 and 3 there was still no provision which allowed any adjustment of wage differentials to deal with labour shortages. This was the first long-term incomes policy to make no provision for such adjustment. Instead, criteria were developed to allow exceptional treatment in a number of circumstances where the underlying rationale was obscure – flexibility, unsocial hours, productivity increases which might be quite unrelated to changes in labour effort or working conditions etc. In a market economy these *ad hoc* expedients are of little importance, but as long as wage flexibility is necessary to reallocate labour, an incomes policy must allow the wage structure to reflect fundamental changes in labour market conditions. The most important case for exceptional treatment arises where genuine labour shortages occur and any future, long-term incomes policy should recognise it.

In the event it appears that the pay restrictions did not last long enough to create permanent damage in the areas affected, although certainly in the Aberdeen area the local construction companies were forced to cut back on house-building programmes. Local firms now face the problems of rising wage costs but the longer-term structural changes that will ensue should be to the benefit of the economy, so long as no future incomes policy shows the same lack of understanding of how labour markets operate.

Regional policy

We begin here with a negative point, but one that follows logically from our previous arguments. It is that there is a strong case for continuing to resist the suggestion that has sometimes been put forward (most noticeably by *The Economist*) to take away development area status from areas such as North East Scotland and consequently reduce the level of regional policy assistance. This view is reminiscent of the policies that operated from 1960 to 1966 under the Local Employment Act. This act gave the Board of Trade powers to provide regional assistance in designated 'Development Districts' which could be redefined on the basis of certain criteria. In practice the Board adopted an unemployment rate of $4\frac{1}{2}$ per cent and scheduled and descheduled areas whose unemployment rate rose above or fell below this level. This had two main disadvantages: the great uncertainty generated for any developer considering investment in a development district (both prospective and existing); and the unsuitability of using unemployment rates as a measure of an area's economic potential.[9]

A system that removed regional incentives from those parts of a development area that have achieved growth is superficially attractive, but it implies that unemployment should always be dealt with in each and every locality in which it occurs, without recourse to labour mobility. Instead we should recognise that a solution to Scotland's economic difficulties will require significant geographical shifts in employment and population. If natural growth centres emerge these can be discouraged only at the cost of creating inefficiency and thus creating a higher general level of unemployment. A regional policy that provides incentives that only take account of unemployment levels will pin its hopes on those areas which generally are the least attractive and offer little possibility of rapid economic growth. This is not a suggestion for positive discrimination in favour of growth areas, but it is a plea to resist the temptation to discriminate *against* these centres should they emerge.

There also appears to be a good deal of misunderstanding about the role played by regional policy in determining the location of North Sea oil activities. Much of that employment is in service industries which do not attract development area assistance. In many areas the chief function of development area assistance is not to benefit new employers entering the service sector, but to assist traditional manufacturing establishments who have to adapt to a situation of extreme labour shortage by higher investment aimed at improving labour productivity. If such assistance was removed, the chief result would probably be job destruction in these firms rather than job transfers between areas. There are, of course, some types of oil-related development, for example production platform activity, which does benefit from development area assistance, and this may be difficult to justify, but this is often a small part of total employment.

Our conclusion would be that whatever regional incentives are available should be applied to all locations within the development area concerned, and we consider the 1974 decision to confer development area status on Edinburgh is a welcome move in the right direction. Yet while the spatial application of regional incentives now appears admirable, the *form* of the incentives may require major modification. Despite the doubling of the Regional Employment Premium, announced in 1974, the incentives available remain heavily biased in favour of capital-intensive development in manufacturing industry; i.e., the subsidy to new investment through regional development grants is a higher proportion of capital costs than the subsidy to wage costs through the Regional Employment Premium. Moreover, while the wage subsidy per job is firmly fixed, regional development grants are open-ended with no limit to the amount of government assistance that can go to any one development. An investigation of how we arrived at a policy that encourages capital-intensive techniques in development areas which allegedly have a surplus of labour and labour-intensive techniques in non-development areas which experience labour shortages would make an interesting thesis, but it evidently owes nothing to first principles.

In Scotland a great deal of the employment created by North Sea developments is in service activities, which do not qualify for regional assistance, and in these areas the chief function of regional policy is not to benefit incoming firms, but to help indigenous manufacturing industry to adapt to a more competitive labour market with higher wage costs. Yet there will be exceptions when manufacturing employment related to North Sea activities is created, and when this occurs the processes will be very capital-intensive and, under present arrangements, will automatically qualify for substantial grants. For example, the proposed oil refinery at Nigg in Easter Ross would have a capital cost of some £150m and provide 350 jobs; with a 20 per cent regional development grant the subvention from the Exchequer would be £30m or £86,000 per job created!

We do not intend to imply that such an oil refinery development is necessarily undesirable, but we do take the view that such a level of assistance is quite inappropriate given the small amount of employment creation. There is then an overwhelming case, on grounds of principle and practice, to recast the form of regional incentives which, as presently constituted, must lead to wasteful and inappropriate expenditure. Two alternatives present themselves immediately. First, a limit could be set to the amount of development grant assistance per job created or, second, the present edifice could be entirely swept away and replaced by a form of assistance which is more factor-neutral as between capital and labour. One method of achieving this, which has been discussed in detail elsewhere,[10] would be to vary corporation tax according to the distribution of a firm's employment between development and non-development areas. There is therefore a case for re-examining the role of regional policy in relation to North Sea developments in Scotland, but made with a view to applying expenditure more effectively rather than to eliminating localities from the receipt of those incentives that the policy determines.

7.7 CONCLUSIONS

The pace, scale and unpredictability of North Sea developments have inevitably created major problems for economic management and physical planning on land. The pressing UK balance of payments need has dictated attempts to encourage much more rapid development of the UK North Sea than, say, the Norwegian sector. This is perfectly understandable given the UK context, but it has had the inevitable consequence of exposing many small communities to economic changes quite foreign to their past experience.

We have argued, in this and the preceding chapter, that the bulk of employment arising from North Sea activities will continue, in the future, to be concentrated in the North of Scotland. There will then continue to be a pressing need to divert more resources to the provision of the necessary infrastructure that these developments will require. Indeed, we take the view that there is little that can be done to disperse such employment creation more widely and, indeed, that an active pursuit of such a policy might be positively harmful to the establishment of a viable and a functionally competitive off-shore industry. The objective can be achieved only if the central government assists in the provision of the finance and other support that will be necessary to relieve local authorities in the North of Scotland from otherwise intolerable burdens, particularly with regard to the provision of infrastructure.

In other areas, too, it will be necessary to modify old attitudes – particularly with regard to regional policy – but above all what is required is the realisation that government policy on-shore has to be formulated and applied in such a way that flexible responses are possible to changing circumstances and peculiar needs. The dictates of the balance of payments have, not surprisingly, been paramount in the early stages of development, and one consequence has been that smaller communities have had to struggle to keep abreast of changes dictated by wider needs. However, on occasion the wider need must also make concessions – if not, the social costs of the economic growth will be very heavy and the pace of North Sea developments, at least with regard to production, will itself continue to be slowed by the delays and bottlenecks that have arisen so often in the past.

NOTES TO CHAPTER 7

1. *North Sea Oil and Gas: Coastal Planning Guidelines* Scottish Development Department (1974).
2. *Study for the promotion of an industrial development pole in Southern Italy* European Economic Community (1966) p. 8.
3. 35 to 40 by 1980.
4. This course would be made more difficult by the fact that the existing law of compulsory purchase forces a local authority to pay development value and not value in current use. Evidently, this would leave the local authority little scope for profit through auctioning the site between competitive bidders.
5. For a clear description of what has already been done to assist these authorities to adapt to new demands see Dr. G. McCrone 'The Role of Central and Local Authorities' Paper read to *The Institute of Petroleum Conference (Aviemore, May 1974).*
6. Average hourly male earnings in the Highlands and Islands were 93 per cent of the Scottish average and 90 per cent of the British average; the corresponding figures for the North East were 89 per cent and 87 per cent. See *New Earnings Survey* London, HMSO (1971) p. 133.
7. See, for example, D. I. MacKay *et al. Labour Markets Under Different Employment Conditions* London, Allen & Unwin (1971).
8. The Pay Board found 'widespread' infringement of the code by established firms in the Aberdeen area (*First Report*, p. 6) and 'in contrast to the allegations, infringements occurred almost exclusively in established firms'. This, of course, was precisely what the 'allegations' predicted. New firms in setting their initial wage level were *not* restrained by this code. Naturally, they found it easier to abide by the code thereafter.
9. These points are discussed in more detail in G. McCrone *Regional Policy in Britain* London, Allen & Unwin (1969) pp. 121–5.
10. D. I. MacKay *A New Approach to Regional Policy* London, Poland Street Paper No. 1 (1973).

CHAPTER 8

The Political Economy of North Sea Oil

There were ten girls, who took their lamps and went out to meet the bridegroom. Five of them were foolish and five prudent; when the foolish ones took their lamps, they took no oil with them, but the others took flasks of oil with their lamps. As the bridegroom was late in coming they all dozed off to sleep. But at midnight a cry was heard: 'Here is the bridegroom! Come out to meet him'. With that the girls all got up and trimmed their lamps. The foolish said to the prudent, 'Our lamps are going out; give us some of your oil'. 'No', they said, 'there will never be enough for us both. You had better go to the shop and buy some for yourselves' ... 'Keep awake then; for you never know the day or the hour'. (Matt. 25, vv. 1–13)

8.1 INTRODUCTION

The distribution of the economic benefits from North Sea oil will depend on the political framework in which they are set and on the priorities and preferences of those who manipulate that framework. A similar comment could be made of any industry, but it has a special force when applied to North Sea oil. The difference is so substantial that it amounts to a difference in kind rather than of degree, and an understanding of the nature of this difference is vital if we are to perceive correctly the nature of the benefits that North Sea oil will bring.

The economic significance of the North Sea oil and gas discoveries lies in the major improvement they will bring to the UK balance of payments. By the later 1970s this will obviate the need for heavy

160

borrowing to finance oil imports and the painful adjustment in Britain's living standards which such borrowing could only delay. The chief immediate beneficiary will not be the consumer in terms of lower energy prices, but the government in the form of oil revenues from royalties and taxation. These revenues will be truly stupendous, as we have seen in chapter 5, and if the government in the expenditure of these revenues does not discriminate in favour of particular regions, then all UK regions will share the benefits in the same degree.

In a very real sense this is the nub of the problem. It may seem just and equitable that these benefits should accrue to all regions in equal measure. Unfortunately, at least unfortunately for this state of apparent harmony, an increasing number of Scots have adopted a rather proprietary view of the oil found off the Scottish coast. This being so, they are likely to judge economic policy towards North Sea oil not only, and not even mainly, by the effects on the UK balance of payments and on government revenue account, but more importantly by the extent to which it removes the disadvantages, low incomes, high unemployment and high emigration rate, which they have experienced unremittingly over the last half-century.

In this chapter we shall investigate the economic consequences for Scotland that would flow from different political and constitutional arrangements. This takes us into contentious territory, but we believe the issues must be stated clearly and frankly. Indeed, they cannot be avoided, for what is now conjecture will ' become reality if our earlier analysis is accepted. If the *economic* conclusions drawn here are mistaken, then the fault lies somewhere in the earlier chapters, for the conclusions follow logically from that earlier analysis.

We distinguish between two effects of oil and gas activities – the employment and income effects arising from North Sea activities themselves (which will be concentrated mainly in Scotland) and the effects on the balance of payments and on government revenue from oil exports (which will accrue, in the absence of special measures, to all regions equally). As we are concerned particularly with Scotland we shall label these as the 'direct' and 'indirect' benefits.[1]

Our subsequent argument can be summarised as follows. There is little prospect that the direct impacts will be large enough to produce a major and lasting change in Scotland's economic performance. There certainly has been a clearly discernible improvement in Scotland's performance relative to other regions of the UK, and this improvement will continue to be observable over the years that lie immediately ahead. However, it seems unlikely to be sufficient to eradicate the historically higher levels of unemployment and lower incomes that Scotland has experienced relative to the remainder of Britain. Indeed, to believe that the direct income and employment effects from oil are the most important benefits is entirely to misunderstand the nature of the industry. The crucial benefits are quite separate from this and of a quite different order of magnitude. They are the balance of payments effects which will largely accrue to the State in the form of royalties and tax revenue and which will allow a more expansionary economic policy than could otherwise be contemplated.

These revenues will place major discretionary power in the hands of the central government and the ultimate benefits of these revenues will be determined by political arrangements and priorities, rather than by strictly economic considerations. The questions that then arise, and which we must now pursue, are fundamentally questions of political economy rather than of economic science. The issues involved are complex, but it is useful to distinguish between the economic consequences of three alternative political and constitutional arrangements – the continuation of 'present policies', the creation of a separate or independent Scotland, and the options that lie between these polar positions.

8.2   PRESENT POLICIES

By 'present policies' we mean a situation in which the major direct benefits of North Sea activities accrue to Scotland, but no further steps are taken, such as the hypothecation of a large portion of the oil revenues for Scottish economic development. It could justly be claimed that no major political party advocates this position, as each has recognised the need to assist Scottish economic development in some such fashion. However, these commitments as yet are vague rather than specific and do not appear to amount to a radical change in present economic relationships. Present policies may therefore be a straw man of our own creation, but it is sufficiently close to reality to provide a starting point for our discussion.

In chapter 6 we provided detailed estimates of the employment that is likely to arise from North Sea oil and gas activities in Scotland. For a number of reasons, which are inseparable from this type of exercise, they must be subject to some margin of error. We cannot even be certain as to the present level of job creation, and if it is difficult to know precisely where we are at present it is even more difficult to predict the future. Taking our courage in both hands we have forecast that direct employment creation in North Sea oil and gas activities will reach a plateau of some 30,000–35,000 before declining from the late 1970s.[2] However, if the underlying assumptions are falsified, and some of them certainly will be, the amount and the timing of employment creation may be different from our forecast.

Primary income and job creation in North Sea and related activities will also have secondary effects on income and employment. This is the normal Keynesian multiplier process – higher incomes in one sector lead to higher consumption and thus to increased demand and increased employment in other sectors of the economy which are not directly related to the primary sector. The size of the multiplier depends on a variety of factors, but for a region such as Scotland it is unlikely to be greater than two.[3] This means that secondary employment will be created on a 1 : 1 ratio with primary employment, so that the total number of jobs arising directly and indirectly through the multiplier process will be in the region of 60,000–70,000.

The process we have described is a very mechanical one, based on employment that has been created or can be fairly readily foreseen, and then going through the usual arithmetic of the Keynesian income on employment multiplier. None of this does real justice to the complexity of North Sea developments, nor to the growth dynamic that they might create. North Sea oil and gas and its assiciated developments will provide a missing stimulus, a new industrial base to sustain expansion. But if so, we are talking about what Hicks in the context of the theory of fluctuations,[4] and Wilson, in the context of regional development,[5] called the 'super-multiplier': something more complex, more dynamic and certainly more unpredictable than the simple and mechanistic Keynesian multiplier. Some process like this is certainly at work in those areas of Scotland that have been particularly affected by North Sea oil impacts. It is evident that in a very short space of time business expectations as to the future have been transformed, and this may well influence a whole range of activities which have little direct connection with North Sea oil. As economists we can do relatively little to analyse or quantify these effects, but we should be prepared to recognise that they exist, for they have significant implications for economic planning and economic management.

We have then to allow for the possibility that there may be favourable effects from North Sea oil activities which we cannot easily allow for in our estimate that job creation, both primary and secondary, will amount to 60,000–70,000. Moreover, the North Sea is a testing ground for what amounts to virtually a new industry in off-shore exploration and production and hopefully may provide Scottish and the UK industry with a technological lead that can be applied in other off-shore areas. It is the proper aim of government policy to assist this process wherever possible. However, economic policy cannot rest on faith and hope alone, and there are good grounds for supposing that the industrial and technological spin-off from North Sea oil may be much more restricted than is often supposed.

A realistic appraisal shows that much of the present employment creation is directly dependent on continued activity in the North Sea. Exploration activity seems likely to peak in 1976 and then to begin to decline slowly. Other types of activity in which employment is growing rapidly at present, particularly the construction of oil production platforms, also seem liable to reach a peak in the late 1970s. Employment is then likely to decline more quickly than during exploration and may create very substantial unemployment in particular labour markets. Similarly, there will be a substantial increase in employment as a consequence of the construction of production facilities, tanker terminals, gas terminals, oil refineries and associated investment, in the mid- and late 1970s, but this too, in the nature of things, must be a peak from which employment will decline.

Most of the oil-related employment in Scotland is dependent on the level of activity in the North Sea, and indeed employment in exploration and the construction of production facilities is *entirely* dependent on North Sea activities. When this activity shifts to other off-shore areas employment in Scotland must decline. It seems quite clear that employment in the oil-producing phase is likely to be significantly below employment in the exploration stage and during the manufacture and construction of production facilities. Finally, some types of activities that occur during the production phase may have much less impact than many hope and expect. Thus we have argued previously that while some additional refinery capacity may be located in Scotland, it is likely to be limited and have little spin-off. Forceful government intervention may change the situation, but in any event refineries are very capital-intensive and employ extremely little labour.

Hence, while our estimates of employment creation may well prove to be conservative, we do not believe that the eventual outcome is likely to be radically different from that forecast above. At least this appears to be the prudent assumption on which economic policy should be based. Given this, we have to consider whether the employment creation that we have forecast is likely to bring a major change in Scotland's economic performance.

The basic economic problem in Scotland is easily stated – the need and the difficulty in establishing a new propulsive sector, a new industrial and commercial base to replace the dynamic once imparted by traditional industries. The attempts to build up alternatives, such as motorcar manufacturing and a light engineering base centred on electronics, have been important in providing new employment opportunities, but they have had to swim against a continuing ebb tide of declining employment in the traditional industries. The results are well known – since 1920 Scotland has, next to Northern Ireland, been the most disadvantaged region of the United Kingdom. Unemployment has always been significantly above the British level; employment incomes have been some 8 per cent below the British average, with relative *per capita* incomes lower still; and the rate of emigration has probably been higher than that from any other comparable industrial region in Europe.

In the postwar period these discrepancies were most marked in the early 1960s, but subsequently, and largely as a result of the more active regional policy pursued from 1963, Scotland's relative economic performance has shown some improvement. The unemployment relative has fallen; employment and *per capita* income have crept up a little towards the British average; and net emigration has declined. In view of this the recent judgement that regional policy has been 'a distinguished failure'[6] may appear a harsh judgement, but it does reflect the difficulty of offsetting the sheer weight of declining sectors.

For analytical purposes it is useful to distinguish the 'propulsive' or 'export base' from other sectors of a regional economy. The former consists largely of primary, extractive and manufacturing industries on which employment and income in the tertiary sector, construction and service industries, largely depend.[7] Over 1966–72 employment declined in every primary, extractive and manufacturing sector save leather, clothing and footwear, where there was a modest employment gain.[8] In the basic sector as a whole employment declined from 888,000 in 1966 to 734,000 in 1972, a fall of 154,000 or more than 17 per cent. To put it another way, the basic sector was losing jobs at a rate of 25,000 per annum over the six-year period. Against this we have to set the secular tendency for employment to increase in the tertiary sector so that over 1966–72 total employment fell more slowly than employment in the propulsive industries. Nonetheless, the employment gains were not sufficient to offset the drag on the economy of the declining basic sectors, which was aggravated by a sharp fall in employment in the construction industry over this period.

This phenomenon of job losses in the basic industries is not merely a product of the last few years. Particularly in the case of males, employment in these industries has been declining over a longer period. Nor is there any indication that the process is at an end. The job losses have been heavily concentrated in West Central Scotland and the increasing economic difficulties of this area have been a feature of the 1960s. The magnitude of the task is best summarised by the detailed official report recently completed for the West Central Scotland Planning Committtee:

> Over the period since 1960, West Central Scotland has had an unemployment rate consistently higher than in any other region of Great Britain, running at over twice the national average for most of the time and being exceptionally severe for males . . . the level of net emigration . . . has also been exceptionally high, running at 24,000 per annum during the 1960s, a rate considerably in excess of any other region in the United Kingdom . . . both manufacturing and total employment declined between 1959 and 1968, by 5% and 1% respectively, a worse record than in any other region of the UK.[9]

Other indicators – low *per capita* income, low labour force activity rates and greater vulnerability to cyclical fluctuations – show the same dismal picture, but the most graphic illustration of the extent of the problem lies in the comparisons made between the future job requirements of West Central Scotland and those of other areas of the European Economic Community. West Central Scotland job requirements over 1971–80 were estimated to lie between a minimum of 153,000 and a maximum of 236,000, and even the latter figure assumed a continued loss of 10,000 persons a year through net emigration. For the declining regions of France, Italy, Holland and Belgium, job requirements, estimated as far as possible on a similar basis, were 150,000, 100,000, 80,000 and 120,000 respectively. Hence, the *minimum* requirements of West Central Scotland, which is only one part of a small economy, lie above the total requirements of all the declining regions of far larger economies such as France and Italy.[10]

This is the backcloth against which the income and employment creation arising directly from North Sea activities must be viewed. The activities present a major new dynamic, new propulsive sector to set against the decline of the older industries. The psychological effect may be as important as any other, for hope is offered where previously there was only despair. Government policy should be directed to building on this foundation wherever possible, but it must also recognise, if it is to be successful, that, viewed against the extent of the problem, there are distinct limits to the regeneration that can be accomplished by the direct impacts which we have analysed. What we have observed over the early 1970s is an improvement in the economic performance of Scotland as the flood tide of rising North Sea employment has offset the ebb tide of falling employment in traditional industries. In consequence, this was a distinct improvement on the unemployment relative in 1974, and over 1973–4 immigration to Scotland exceeded emigration for only the third time in the twentieth century. North Sea employment will continue on a rising trend in the immediate future but nothing is more misleading in forcasting than the simple assumption that a trend, once established, will continue indefinitely. No trend ever does. While we cannot be certain as to timing, we must expect that the rate of growth of North Sea employment will eventually slacken and that this will be followed by some downturn in employment.

Yet, there are other encouraging factors. The incentives provided by regional policy to development areas such as Scotland remain more generous than those of any previous period. Moreover, the stronger balance of payments position provided by North Sea oil will permit more expansionary UK demand management policies, and the traditional wisdom has it that development areas do particularly well in a period of more rapid growth.[11] These effects, combined with the stimulus of the direct impact of North Sea oil activities, *could* be sufficient to bring about a sea-change in Scottish economic prospects. However, we consider this unlikely, or at least a very high-risk strategy. At the peak, direct job creation in North Sea activities will offset less than two years' decline of employment in the propulsive industries, if that decline continues at the rate experienced in the late 1960s and early 1970s. Total job creation from North Sea activities, including all primary and secondary effects, is at its estimated peak of 60,000–70,000 jobs equivalent to less than one-half of the *lowest* estimate of job requirements for West Central Scotland over 1971–80 and to only 3½ per cent of Scotland's employed labour force in 1973. Moreover, the peak of employment in oil and its related industries will not be maintained for long and from the late 1970s employment is likely to decline. Viewed in this light it is difficult to avoid the conclusion that North Sea activities offer too narrow an industrial and commercial base to eliminate the depressingly familiar Scottish symptoms of low incomes and high unemployment. There will be an improvement in the health of the patient, and some signs of this are already apparent, but he will require more drastic surgery to make a complete recovery.

8.3   THE ECONOMICS OF INDEPENDENCE

Here we are considering a quite hypothetical situation, but one in which an increasing number of Scots are taking a real interest. There can be little doubt that the discovery of North Sea oil and the rise in the Scottish National Party are not unconnected events! The nationalists argue that an independent Scotland would acquire sovereignty over the oil and gas fields in the middle and north North Sea and that the extent of the discoveries would guarantee a strong balance of payments and a more rapid rate of economic growth.

The legal position need not detain us long. It is more complex than is often realised, but it is difficult to deny the proposition that, given the terms of the Geneva Convention, an independent Scotland would acquire the right to exploit the oil and gas fields thus far established. As we have seen,[12] the Convention provides that signatories should first attempt to establish their respective jurisdictions by agreement. Failing this, the principle of equidistance is to be applied 'unless another boundary line is justified by special circumstances'. This principle would establish a Scottish sector smaller than the area of 'Scottish' waters defined under the Continental Shelf (Jurisdictional) Order of 1968, which draws the boundary along the 55° 50′

parallel. Instead, the boundary would run north-east from the Scottish–English border, but it would pass south of the most southern oil field yet discovered. Even if the principle of equidistance was modified, it would seem that this could affect only the small southerly fields of Argyll, Auk and Josephine. For these reasons we have conducted our analysis on the assumption that an independent Scotland would acquire the right to all the oil and gas discoveries in the middle and north North Sea.

It might be useful to begin our discussion of the second polar position by considering the following question, for the answer to the question indicates the true economic and political significance of North Sea oil: why is it that the anticipated exports of North Sea oil will yield huge government revenues and the possibility of transforming the balance of payments of the large UK economy, while the direct effects, accruing mainly to Scotland, appear likely to leave a far smaller economy with relatively low incomes and high unemployment?

Let us first eliminate two explanations that are sometimes put forward. The fundamental reason for this paradox does not lie in the capital intensity of North Sea activities, or in the import content of the equipment and services used in the North Sea. It is true that North Sea activities *are* very capital-intensive, and that the import content of equipment and services *is* high, but much more important is the fact that the resource cost of North Sea oil production is very low relative to its market value. As we saw in chapter 2, the owner of the natural resource can exact a high economic rent, for the value added in North Sea is high relative to the expenditures necessary to obtain production. It is the volume of these expenditures that will determine the limit to the direct employment and income effects on Scotland, while the import content of goods and services will determine how close we approach to this limit. In chapter 5 we calculated that the total capital and operating expenditures per barrel of production would amount to $2.26, $2.43, $1.12 and $1.88 for the Argyll, Auk, Piper and Forties fields respectively.[13] These expenditures would set the upper limit to direct income and employment creation in Scotland *even if all the direct expenditures accrued to Scottish concerns.*

The market value of the oil produced is however far greater, at some $11 a barrel. We have shown that a net return, after tax and royalties, of $4 a barrel would be likely to yield a taxation regime which would allow private companies a high rate of return on capital

invested in North Sea activities. There is no escaping the conclusion
that the resource cost of oil production, and therefore the direct
employment and income creation, is low relative to its market price
and that therefore the direct effects must be much smaller than the
benefits that will accrue through the balance of payments and to the
government through royalties and taxation.

We can, indeed, obtain a fairly clear idea of the extent of this
difference. The taxation regime we have suggested would yield, after
the first two years' production life of most fields, a government
revenue (in royalties and taxes) of $7 a barrel. The total capital and
operating costs per barrel are unlikely to be greater than $2 a barrel,
as the larger fields all seem to have costs below this level. Very
approximately, we can take this as being evenly divided between
capital and operating costs. At present, it seems that only some 35
per cent of capital expenditures accrue to UK firms.[14] The share of
the market taken by firms operating in Scotland must be lower – let
us say, very optimistically, 30 per cent. The import content of operating expenditures is considerably lower and we shall assume, again
optimistically, that 70 per cent of all these expenditures accrues to
companies operating in Scotland.[15] This would suggest that of the
market value of $11 a barrel at the most $1 will accrue directly to
companies operating in Scotland (30 per cent of capital expenditures
at $1 a barrel and 70 per cent of operating expenditures at $1 a
barrel), while the revenue accruing to the central government will
be some *seven* times higher!

There is nothing unique about North Sea oil in this respect. Indeed,
as its costs of production are relatively high, the direct income and
employment effects resulting from a given volume of production
will be greater than those which arise in low-cost regions of production. However, a multiple of a small number often remains a small
number! So it is in this instance. The ratio of direct income and
employment effects to the market value of final output for North
Sea activities will be higher than in, say, the Arab oil-producing
nations, but it will remain a small ratio. The revenue from oil will
be of a quite different order of economic importance. It is the
revenues from oil and not the direct income and employment effects
that make the desert bloom, and to this the North Sea will be no
exception.

The direct employment and income effects on a Scotland with
economic and political independence would be no greater than those

discussed in the previous section. On the contrary, the direct effects would probably be smaller in the short run as an independent Scotland would be likely to aim for a much lower level of oil output than that presently contemplated. However, oil exports would produce an extremely strong balance of payments position and a high level of government revenue which would permit, and indeed demand, the pursuit of a very expansionary economic policy. These 'indirect effects', as we have called them, are of crucial importance.

The magnitude of these effects can be calculated by adapting the estimates of chapter 5. We have already discussed the taxation regime that might be applied and we simply assume the regime applied in chapter 5 which yields estimates of government revenue as shown in table 5.4. To estimate the 'oil' balance of payments for an independent Scotland does, however, require some further assumptions. These are:

1. Oil output is as shown in table 5.1, but we value the oil at $11 a barrel, and not at $11.50 as for UK exports, as we assume that the oil would not be exported in ships of Scottish origin. The value of exports is therefore equivalent to total sales revenue as shown in table 5.4.

2. Scotland's oil consumption is only some 10 per cent of the UK total.[16] We substitute in table 5.6 the value of Scottish oil imports calculated as one-tenth[17] of the UK estimated requirements over 1974–80.

3. In 1974, 25 per cent of the goods and services required for North Sea oil developments were supplied by Scottish firms and this proportion will rise by 1 per cent per annum to reach 30 per cent in 1979–80 (this compares with an assumption of a 35 per cent share for UK firms rising to 47 per cent in 1980). The remainder represents imports and when applied to expenditures on exploration and oil development in table 4.3 yields the estimates shown below.

4. Seventy per cent of operating expenditures will accrue directly to firms established in Scotland, so that the remaining 30 per cent represents imports. The latter figure is applied to operating expenditures as estimated on p. 102.

5. Ten per cent of the capital and operating expenditure on North Sea oil and gas is financed from Scottish domestic sources (compared with 30 per cent in the UK case). The remainder of the expenditure required to finance exploration and oil development expenditure in table 4.3 then represents the rate of inward investment.

6. Ninety per cent of profits and interest from North Sea opera-

tions is remitted overseas (compared with 55 per cent in the UK case). This is then applied to estimated company profits as calculated on p. 102.

TABLE 8.1 'OIL' BALANCE OF PAYMENTS OF AN INDEPENDENT SCOTLAND, 1974–80
(£m)

	1974	1975	1976	1977	1978	1979	1980
1. Exports (oil)		209	986	2095	3186	4818	5534
2. Imports (oil)	336	339	342	339	339	355	368
3. Imports of equipment and services	563	759	876	810	728	630	543
4. Import content of operating expenditures	*	6	27	57	87	131	151
5. Visible trade balance (1) − [(2) + (3) + (4)]	− 899	− 895	− 259	+ 889	+ 2032	+ 3702	+ 4472
6. Invisibles: profit and interest remitted overseas	0	148	696	896	1304	1725	1519
7. Current balance (1) − [(2) + (3) + (4) + (6)]	− 899	− 1043	− 955	− 7	+ 728	+ 1977	+ 2953
8. Inward investment	675	923	1080	1013	923	810	698
9. Balance of current and capital accounts (1) + (8) − [(2) + (3) + (4) + (6)]	− 224	− 120	+ 125	+ 1006	+ 1651	+ 2787	+ 3651

*negligible

Summarising the results, the oil balance of payments for an independent Scotland is shown in table 8.1. The government revenue would be the same as that calculated for the UK in table 5.4[18] namely:

	1975	1976	1977	1978	1979	1980
Total government take (£m)	26	123	909	1447	2463	3343

As Scotland's need for imports is so small relative to the UK the oil balance of payments could have a small surplus by 1976, and that surplus would rise rapidly to reach an estimated £3651m by 1980. To appreciate the magnitude of these effects on an independent Scotland we have to set the above estimates against Scottish national income. In 1972 Scottish gross domestic product (GDP) at factor cost was 8.6 per cent of the United Kingdom total.[19] No information for Scotland is available after that date, but the relationship with UK domestic income has changed very little over a number of years. UK gross domestic product at factor cost was £62,176m in 1973, so that Scottish GDP will have been in the region of £5350m. The arithmetic is now virtually complete and the results are startling. By 1976 the net surplus on the oil balance of payments of an independent Scotland (measured in 1974 prices) might amount to some 2 per cent of 1973 GDP; by 1978 to 31 per cent and by 1980 to 68 per cent of total 1973 GDP. Government revenues from the sale of oil abroad would be equivalent to 2 per cent of 1973 GDP by 1976, 27 per cent by 1978 and 62 per cent by 1980.

These results are obtained by employing what we believe to be the most reasonable assumptions given current knowledge, but the reader should be quite clear as to the status of the estimates. In the nature of things they must be subject to some margin of error and the error in either direction could be quite large. Any analysis that rests on these particular figures being realised must, therefore, be subject to considerable reservations. However, the essence of the argument that we deploy below does *not* depend on these particular targets being met; it would hold even on a much more conservative set of assumptions. Thus, if we assume, as in chapter 5, that the price of crude might fall to $7 a barrel for imported crude and $8 a barrel for North Sea crude, then this would still yield by 1980 a surplus of some £2550m on the oil balance of payments of an independent Scotland and government revenue in the region of £2000m. Hence, even this pessimistic assumption cannot shake the conclusion that North Sea oil would provide an independent Scotland with a very strong balance of payments position and substantial surplus of government revenue over expenditure.

The surplus on oil balance of payments account could be used to build up foreign exchange reserves, but this could be only a temporary position, as adequate reserves would be established very rapidly. Both international obligations and internal needs would then point in the same direction – to take action to offset the surplus on current account arising from oil. At the ruling exchange rate this could be done by:

1. following an expansionary domestic policy by cutting taxation and/or increasing public expenditure, which would lead, through the multiplier process, to higher employment and incomes. As incomes rose and the capacity output of the domestic economy was approached then imports would tend to increase, until at full employment any further expansion of demand could be met only through additional imports;

2. holding taxation and public expenditure at present levels, using the additional revenue to increase public sector saving and neutralising the effect of this on the domestic economy by investing the revenue in interest-earning assets abroad. The increased export of capital could then offset the surplus on current account, but it would have no effect in the short run on domestic incomes and employment;

3. reducing the volume of oil production, thereby directly reducing the initial surplus on the oil balance of payments.

In practice some combination of these three measures would need to be adopted. The first priority would undoubtedly be the pursuit of a more expansionary economic policy aimed at raising the level of demand and reducing unemployment. A reduction in taxation and/or an increase in public expenditure would lead to an increase in income and employment in the domestic economy and also to a deterioration in the current surplus of the balance of payments as higher incomes sucked in greater imports. The balance between the rise in domestic output and in imports would depend on the absorptive capacity of the domestic economy, and this in turn depends on the view taken as to the causes of the historically high unemployment experienced in Scotland.

To the extent that unemployment is the result of deficient demand, then it can be effectively reduced by more aggressive demand management policies such as those outlined above. In this case domestic output would rise quickly at first as unemployed labour reserves and under-utilised capital equipment were brought into production, and the rise in imports would be correspondingly low. However, as the domestic economy approached full employment a continued expansion of demand would create strong inflationary pressures and imports would rise rapidly. At the same time, exports would become less price-competitive in overseas markets and this tendency would be aggravated if the exchange rate were allowed to float upwards to neutralise the inflationary effects of the oil revenues on the domestic economy. Thus, while an expansionist demand management policy would bring some benefits in the form of higher domestic income and employment, it would, if carried too far, adversely affect the international competitiveness of substantial sections of Scotland's traditional exporting industries.

Moreover, the extent to which unemployment could be reduced and domestic output increased by demand management turns on the underlying causes of the historically higher levels of unemployment experienced in Scotland. While a large current surplus would certainly allow the elimination of unemployment caused by deficient demand, it provides no magic cure for unemployment owing to structural imbalances and an inability to compete effectively with overseas producers. The problem created by an unfavourable industrial structure, with a high proportion of Scotland's resources of physical and human capital tied into industries and products with unfavourable market prospects, would remain as real as ever. Nor would oil revenues provide any guaranteed remedy to low labour productivity and that defensive psychology, so deeply embedded in the Scottish subconsciousness, which has produced such unenviable obstacles to economic growth in housing policy, restrictive labour practices, a rooted aversion to change and a system of industrial relations which sometimes appears to have been laid down in the same era as the oil deposits.

It should be obvious from this that 'Scottish independence and Scottish oil' would not produce, at a stroke, a type of demiparadise flowing with Cadillacs, milk, honey and whisky galore. It would certainly be possible to cut taxes on personal incomes and allow higher consumption demand to spill over into additional imports. The size of the revenues would obviously allow a very substantial increase in imports to be sustained over a long period, and so the standard of living would rise appreciably. Yet, a consumption-led boom of this nature would do nothing to strengthen the ability of export-based industries to increase production. On the contrary, it would almost inevitably weaken their position appreciably and could well produce both higher consumption, with the higher level of imports sustained by oil revenues, *and* higher unemployment.

Even if one takes a relatively optimistic view about the margin of spare capacity in the economy, and hence about the extent to which unemployment can be dealt with by appropriate demand management policies, it is clear that, given the extent of the likely revenues arising from oil, any attempt to absorb them immediately would only lead to severe inflationary pressure. The limit to reducing unemployment through demand management would certainly be reached when the surplus on the current account would still be embarrassingly large. Hence, economic management in an independent Scotland would have to be carefully conducted. The proper policy would be to stimulate demand, not so much by encouraging additional consumption but by stimulating a higher rate of private and public investment. The appropriate instruments might include a lower rate of corporation tax; grants to encourage a higher level of private investment in capital equipment and buildings and also in training; special incentives to encourage direct foreign investment in the economy; and increased public investment expenditure, particularly in those areas and sectors that reduce costs and promote a more efficient utilisation of resources, e.g. in communications generally, in public training and retraining programmes and in types of industrial development, such as those proposed for Hunterston, that require a substantial public commitment before private investment is forthcoming.

However, it is clear that, even given such policies to increase the underlying growth rate of the economy, its absorptive capacity would still be small and would grow slowly relative to the size of the oil revenues. An independent Scotland would then have to be prepared to invest capital abroad, reduce oil production or adopt some combination of these measures. If oil prices were expected to decline the rational policy would be to set a high rate of extraction, neutralising the expansionary effect this would have on the domestic economy by using the resulting revenues to finance capital investment overseas. Such a policy would have the additional advantage that it would reduce the reliance of other oil-importing countries on Middle East producers. On the other hand, the factors that might cause a UK government to limit output (see pp. 43–4) would weigh even more heavily with a Scottish government, particularly the desire to establish a strong indigenous force in the oil industry and to reduce the pressure on those areas directly affected by North Sea activities. This being so, it is unlikely that output would much exceed 100 million tons a year. Even at this level it would need to be accompanied by major investment abroad and a willingness to allow a gradual revaluation of the currency to damp down inflationary pressures in the domestic economy.

The oil revenues would present much the same opportunities, and the accompanying difficulties, to an independent Scotland as those noted for Norway by a recent OECD report. It found:

> The immediate labour market impact[20] of the oil exploration is and will remain relatively small. . . . By far the biggest part of income generated by the oil sector, and which is not directly spent or transferred abroad, will accrue to the Government.[21]

OECD considered that, even given the cautious Norwegian approach to production, the effect of oil revenues would be to convert a current account deficit equivalent to 2 per cent of GDP in 1971–3 to a surplus of 7–8 per cent in 1977, by which date extra revenue would 'be equivalent to more than 13 per cent of estimated total tax receipts in 1974'. Even by this date, when output would be considerably short of the peak level likely to be achieved in the early 1980s, OECD concluded that the absorption capacity of the Norwegian economy would be limited and 'that serious transitional problems can only be avoided if the domestic spending of oil revenues is stepped up rather cautiously'.[22] It should be noted that the ratio of oil reserves to domestic income would be much higher for an independent Scotland than for Norway.

Ownership of the oil revenues would not automatically remove the fundamental causes of Scotland's economic difficulties in the past. Indeed, without the application of sensible economic policies, higher living standards, while the revenue from oil lasted, could significantly weaken the long-run competitive position of the economy. Nonetheless, the oil revenues could bring real and lasting economic benefits. They would provide a large current account surplus on the balance of payments. This would allow a substantial export of capital which, like Britain's investment overseas in the nineteenth century, would give rise to a continuing income in the future in terms of remitted interest and dividends from overseas. Domestically, it would be possible to pursue expansionary demand management policies aimed at eliminating unemployment owing to deficient demand. Some immediate increase in domestic output and employment could then be expected. Within a relatively short period further expansion would depend on longer-run policies designed to improve the efficiency and the competitiveness of the economy. However, it could be argued, and with some force and persuasiveness, that the oil and gas discoveries in the North Sea have created an environment in which these real difficulties might be successfully tackled. Indeed, whatever views might be adopted on other grounds, we find it difficult to deny the proposition that the revenues from oil must fundamentally alter the economic prospects that would be faced by an independent Scotland.

8.4   THE ECONOMICS OF DEVOLUTION

We have drawn a rather bleak contrast between the economic consequences of continuing 'present policies' and of independence. Present policies would retain the union as we have known it, but would be likely to leave Scotland relatively impoverished. Independence would provide Scotland with major benefits for the foreseeable future. Yet, mere economics is not necessarily decisive, and the foreseeable future is a short period set against more than 250 years of union. Many might consider that the political and social costs of independence are too high. We have then to consider whether there are intermediate solutions that lie between these two polar cases.

It may be claimed, with some justification, that our description of 'present policies' created a convenient straw man which would subsequently be demolished. Indeed, each of the three major UK political parties has declared an intention to adopt intermediate solutions which can be grouped under the terms of 'devolution' and 'federalism'. These terms have many shadings and nuances and cover administrative and legislative as well as the economic powers. Of course, these issues are not unconnected, but some distinction can nonetheless be made. Administrative and legislative devolution is important in its own right, but it is largely irrelevant to the central issues of economic management. The devolutionary schemes thus far proposed accept the continuation of the major features of the present system of economic management. This is not to say that the schemes contain no economic proposals, but these proposals do not fundamentally affect the present system wherein the levers of economic power are effectively controlled by the central UK government.

Our concern is with the economic issues, particularly with devising a system that recognises and deals effectively with Scottish economic problems. The discussion of the previous section provides an insight into the nature of the solution which might prove effective. As we saw, the crucial benefit that the oil revenues would bring to an independent Scotland would be the creation of a strong balance of payments position which would allow the pursuit of more expansionary economic policies aimed at increasing the productive capacity of the economy. Any suggested solution that does not approximate this situation and does not permit the pursuit of policies better fitted to Scotland's own economic needs is therefore likely to be ineffective. This is the critical test against which all proposed remedies must be evaluated.

It would seem to us that all the schemes thus far proposed under the heading of devolution fail this critical test, as they do not change the basic framework within which UK economic management has been conducted. For the most part, devolution is concerned with the transfer to a Scottish Assembly of legislative powers in those fields where the Secretary of State for Scotland, through the Scottish Office, already exercises executive and administrative responsibility – housing, health, education, planning, local government etc. Devolution would provide an Assembly, through a block grant determined by Westminster, with financial discretion over the distribution of expenditure between these different functions. This is an extremely important political and constitutional development, but the Assembly would operate within a strict upper budgetary limit set by a central government, which would maintain sole control over the level of expenditure.

The proposals for devolution do not change the basic nature of the economic relationships between Scotland and the rest of the UK. This is that the major economic decisions are taken by central government and apply uniformly to the UK regions. Offset against this is regional policy, again determined at the centre, which attempts to raise the level of activity in high unemployment regions through financial incentives to private investment and through the regional employment premium and other measures intended to lower costs of production. In addition some regions, including Scotland, have obtained in the past a higher proportion of public investment than would be justified on a *per capita* basis[23] and in recent years more strenuous attempts have been made to redistribute public sector employment in favour of the slow-growing regions. The White Paper published by the last Labour administration appears to envisage a continuation of these policies, although they would be pursued more vigorously.

The most important new departure would be the creation of a Scottish Development Agency, 'which is to be responsible to the Secretary of State for Scotland. Its task will be to strengthen the instruments available for promoting the development and restructuring of the Scottish economy.'[24] In other words, the Scottish Development Agency will allow the pursuit of a somewhat stronger version of present regional policy. It would appear that the agency would be charged with restructuring Scottish industry through the provision of loan and equity finance and possibly through more direct forms of intervention.

Past experience does not suggest that such an approach will be successful. Regional policy operates on a very narrow front, mainly through attempting to raise the level of private investment. The incentives currently available are already generous and it is difficult to believe that further incentives could have any major effect. While regional policy has certainly had its successes, it works only slowly and uncertainly. Indeed, nothing else can be expected, for as long as regional policy operates on a narrow front it must always be less important than the full range of the macroeconomic policies that are deployed by the central government on the basis of UK considerations. Regional policy can at best only mitigate the effects of demand management policies which are applied to deal with UK problems rather than with the problems of particular regions.

For this reason we do not believe that devolution, as currently proposed, will bring a major change in Scotland's economic prospects, for it would leave intact the present system of economic management. Against this there would be some strengthening of regional policy (through the Scottish Development Agency), a certain amount of additional public investment and the transfer of certain administrative functions, such as the British National Oil Corporation, to Scotland. The shadow is more impressive than the substance. The location of administrative functions, such as the BNOC, is less important than the objectives that determine policies. Regional policy would have to continue its uphill fight against the mainstream of economic management dictated by UK considerations. For the most part it would continue to be unable to insulate Scotland effectively against deflationary policies dictated by the needs of the UK economy.

In maintaining a system of strong central control over the management of the economy, the current devolutionary proposals are consistent with the spirit and intentions of the Majority Report of the

Royal Commission on the Constitution.[25] Hence, the Scottish Assembly will have discretion in allocating public expenditure between different uses, but all ultimate economic sovereignty is retained at the centre, which determines the limit to that expenditure. However, even the Majority Report considered that it was important that such an Assembly should have some source of finance *independent* of Westminster,[26] and the Minority Report made more radical suggestions in this direction.[27] The current proposals for devolution provide no independent source of income to the Assembly and this is a fundamental limitation of its powers. Even less do they represent a concession to the ingenious financial and economic scheme suggested by the Minority Report, which proposed that 'intermediate level governments' should have (1) major independent sources of revenue and (2) considerable discretion over the allocation of their expenditures, within a system specifically designed to allow them greater power to eliminate regional disparities of unemployment and incomes.[28]

It is quite clear from the experience of other countries that 'fiscal federalism', which provides state or provincial governments with clearly established, constitutional powers to raise revenue independent of the federal authority, can be consistent with effective stabilisation by the federal government. Yet the Treasury view, that this is inconsistent with effective management of the economy, has prevailed. The comment of the Minority Report of the Royal Commission remains the most apposite:

> However, this line of argument prompts one to ask the question why it happens to be the case that major countries with devolved systems of government manage to be at least as successful (and in some cases more successful) at stabilising their economies with high rates of employment and a lower rate of inflation and, furthermore, how far one can accept the conclusion of various fiscal experts that our budgetary policy in the 1960s de-stabilised rather than stabilised the economy.[29]

Australia, Canada, West Germany and the United States, each of whom can claim a stability and a growth record significantly superior to the UK, operate systems wherein the central government's control over public expenditure is limited by the fiscal autonomy of state governments. The central authority retains the ultimate responsibility for stabilisation, but it attempts 'to balance central and private demand *given* regional taxation and expenditure policies'.[30]

Fiscal autonomy and stabilisation are not then incompatible, but would it be possible to provide a Scottish Assembly with sufficient powers to follow a more expansionist economic policy? In most cases the answer would be negative. The central authority must remain in sole control of the money supply and the conventional stabilisation measures of a state or provincial government must therefore depend on tax and expenditure policies. A provincial government wishing to raise the level of demand by deficit financing could then finance the deficit only by borrowing. It is easy enough to show that this must usually involve borrowing from non-residents and the imposition of an external debt on residents. For these reasons it is generally accepted that:

> ... A local government is likely to be severely constrained in its ability to influence the community's level of output and income. Local government can employ only fiscal stabilisation tools, and the problems of openness and external indebtedness seriously impair the freedom and effectiveness with which these tools can be used.[31]

However, to every generalisation there is an exception and so it is in this case. The hypothecation of a substantial part of the oil revenues to a Scottish Assembly with some independent powers over taxation and expenditure would certainly allow it to follow more expansionist economic policies *without* incurring external debts. The attraction of this system is that it would allow the pursuit of policies based on Scottish needs. Its weakness historically was the fact that a Scottish Assembly could not, because of low incomes per head in Scotland, have sustained the same level of public expenditure without resort to borrowing or deficit financing.[32] Hence the room for manoeuvre of a Scottish Assembly would have been limited and it would have had little influence on the standard of living of Scotland relative to the rest of the UK. This argument can no longer apply, or at least it cannot apply to the Scottish Assembly with a substantial, hypothecated share of the oil revenue.

It will be said that such a system is incompatible with the proper management of the UK economy. This should be interpreted to read that it is incompatible with management as it has been practised. In view of the lamentable economic performance of the UK, and of Scotland in particular, this should be regarded as a positive virtue! British economic policy has always begun and ended with short-run management, as befits a system erected almost exclusively on Keynesian principles. No account is taken of resource utilisation in different regions. Hence, fiscal and monetary weapons are deployed on the basis of UK considerations, although they may be inappropriate to the needs of regions with high unemployment.

The system we have proposed would run entirely counter to this philosophy, for it recognises the need to adapt economic management to the needs of different regions. It would be necessary to take account of the spillover between different regions as the regional economies of the UK are very 'open' and an expansionary economic policy in Scotland would be found to have an effect on incomes and employment in other regions of the UK. This would complicate economic management in the UK as a whole. However, the tail does not usually wag the dog and it is difficult to believe that a more expansionary economic policy for the small Scottish economy could have much effect on the rest of the UK.[33] Certainly, the spillover would be likely to be within the margin of error of most economic forecasts. The real problem has never been the effect of a spillover from Scotland to the rest of the UK, but the difficulty, under the present system, of raising the level of activity in Scotland without an expansionary policy in the UK as a whole. Our suggestions could not entirely insulate Scotland from events in the rest of the UK – no policy could. However, they would achieve, much more effectively than current arrangements, greater freedom of action to attack Scottish economic problems on a broad front.

The key to this process is the hypothecation of a significant part of oil revenues to a Scottish Assembly with some fiscal discretion over taxation and expenditure. Here we should underline one of our previous arguments as a counter to a major misunderstanding which has tended to obscure recent discussion. It is often argued that present UK borrowing against the oil deficit 'mortgages' future revenues so that they will have to be used to finance interest payments on past borrowing and will not, therefore, be available for regional economic development or other domestic economic purposes. This view has some validity, but for the most part it is based on a simple misunderstanding of the economic issues involved.

The argument usually runs along the following lines. Over the next few years the UK will have to borrow heavily to finance its oil deficits. Thus, taking the figures in table 5.6 (p. 103), a total borrowing of £11,618m will be necessary to finance the total UK oil deficit to 1978. Assuming a market rate of interest of 10 per cent it will be necessary after 1978 to service interest payments on that debt of some £1200m annually. As can be seen from table 5.6, this would account for almost all of the surplus on the 'oil' balance of payments even as late as 1980. Therefore, the argument runs, there will be very little surplus left over for other purposes.

Now it is true that such a debt has to be serviced, and to this extent some part of the revenues *are* mortgaged, but the major part of the revenues must be spent unless the government *deliberately wishes to engineer a major departure from full employment.* The reason for this is that the rise in the price of crude oil will have, as we have seen, a very deflationary effect on the UK economy. The revenues that will result from the export of North Sea crude will accrue largely to the UK government. It these revenues are not spent public sector saving will rise, but this will not offset the deflationary effects of higher crude oil prices. Given the sums involved, this could only result in substantial unemployment and low income.

To maintain employment and incomes the major part of the revenues *must* be spent, either through a reduction in taxation or through an increase in public expenditure. In this situation major discretionary powers will be placed in the hands of the government and its use of these powers will depend on its economic and social priorities. Thus it could follow demand management policies, which would simply neutralise the effects of the higher price of oil imports so that each region was left in the same relative position as it was in before the rise in crude oil prices. Alternatively, it could deliberately choose to discriminate in favour of regions with low incomes and high unemployment. This is where our priorities would lie. We have been interested mainly in the Scottish case, but the argument that special attention should be paid to Scotland's problems of low incomes and high unemployment can obviously be extended to other regions, such as north England and Wales.

8.5 CONCLUSIONS

The central fact about North Sea oil is that the major benefits arise through the revenues that will accrue to the government and not through direct employment and income creation from North Sea activities. This is as true for the UK as it is for all other oil producers. The control of these revenues, and the use to which they are put, then become of fundamental importance.

The economic, political and constitutional implications of this position are extremely serious and should not be lightly dismissed. It appears unlikely that the direct income and employment effects will be sufficient to bring a major and permanent change in Scottish economic prospects, whatever the rhetoric about the creation of 'offshore capitals' and 'European Houstons'. On the other hand, the indirect effects on the balance of payments and on government revenue are so large that it is impossible to believe that they could not be deployed to alter Scottish economic prospects profoundly.

North Sea oil is quite properly seen by Scots as offering the possibility of a major break from the dreary record of the past. They will judge policy in this light and in no other. This does not necessarily imply that they will choose the naked pursuit of self-interest, but it does imply that they are unlikely to choose any alternative that offers low incomes and poor employment opportunities relative to the remainder of Britain.

Between these polar positions many policies will be paraded for inspection, but the only intermediate position likely to guarantee success is one that brings a major shift in economic and commercial power from the centre, accompanied by fundamental constitutional and political reforms. These will be implacably opposed by all those who believe that every major economic decision must be taken centrally and will not be easily accepted in a conservative society such as Britain with an inherited suspicion of constitutional 'gimmickry'. Yet such a system of sharing oil revenues is alive and well and living in Canada. Nor can the British system of economic management be defended on its record of achievement.

The solution we have proposed would require Scots to modify their claims in the interest of their partners. This may be acceptable if it is accompanied by a genuine attempt to reverse the trends of the last half-century. A minimum requirement for this appears to be the creation of a Scottish Assembly with a large share of oil revenues, some substantial fiscal discretion and executive and legislative control over those functions presently exercised through the Department of Trade and Industry. Such a system would be consistent with the 'essential unity of the United Kingdom'. Nothing short of this is likely to be effective. There can be no guarantee that even these fundamental changes would be enough, but nothing that falls short of such a system seems likely to be sufficient. The future position that we have depicted will be upon us soon. Appropriate action cannot be postponed until tomorrow: the day and the hour is at hand.

NOTES TO CHAPTER 8

1. Many of those experiencing the 'direct benefits' in areas affected by North Sea oil activities might also emphasise that they carry with them substantial economic, social and environmental costs – the same is not true of the 'indirect benefits'.

2. See p. 136.

3. See A. J. Brown *The Framework of Regional Economics in the United Kingdom* Cambridge, Cambridge UP (1972) and D. B. Steele 'Regional Multipliers in Great Britain' *Oxford Economic Papers* (1969).

4. Sir John Hicks *A Contribution to the Theory of Trade Cycle* Oxford, Oxford UP (1950).

5. T. Wilson 'The Regional Multiplier – A Critique' *Oxford Economic Papers* (1968).

6. *Economic Development and Devolution* Edinburgh, Scottish Council Research Institute (1974).

7. The problem of distinguishing the export base is more difficult than this (e.g., how should tourism be treated?) but the simplification does not do undue violence to the facts.

8. See D. I. MacKay and G. A. Mackay *Scotland: A Growth Economy* Edinburgh, Bell, Lawrie, Robertson (1974).

9. *West Central Scotland Plan, Supplementary Report 1, Regional Economy* (1974) pp. 228–9.

10. ibid p. 174.

11. We believe that there is some truth in this view, but the extent of the improvement in development area performance in a period of rapid growth has been exaggerated because the analysis has usually been carried through in terms of *unemployment* relatives only. Examination of *employment* changes shows a much smaller change in relative positions over the trade cycle. See W. Black and D. G. Slattery 'Regional and National Variations in Employment and Unemployment: Northern Ireland, A Case Study' *Scottish Journal of Political Economy,* forthcoming (June 1975).

12. See p. 20.

13. See p. 97. Adjusting upwards (by one-third), the published Wood Mackenzie estimates (June 1974) of capital and operating costs we obtain an estimated cost per barrel of $1.3 for Ninian and the Brent/Dunlin/Thistle complex and $2.3 for Beryl and Montrose.

14. See p. 83.

15. Compared with our assumption (p. 101) that 80 per cent accrues to UK companies.

16. See Department of Trade and Industry *UK Energy Statistics* (1973) p. 63.

17. The actual value of imports for Scotland would probably be *lower* than this calculation suggests as natural gas from the middle and north North Sea would satisfy a higher proportion of Scotland's future energy needs over 1974–80 than would be the case for the UK.

18. See table 5.4, p. 99.

19. The Scottish Office *Scottish Economic Bulletin* (July 1974) p. 9. See also T. L. Johnston, N. K. Buxton and D. Mair *Structure and Growth of the Scottish Economy* London, Collins (1971).

20. i.e. what we have called the direct impact.
21. *Norway* OECD Economic Survey (1974) p. 32.
22. ibid p. 35.
23. See D. N. King *Financial and Economic Aspects of Regionalism and Separatism* Research Paper 10, Commission on the Constitution (1973).
24. *Democracy and Devolution Proposals for Scotland and Wales* Cmnd 5732, London, HMSO (1973) para. 34.
25. See *Royal Commission on the Constitution, 1969–1973, Vol. 1: Report* Cmnd 5460, London, HMSO (1973).
26. ibid. paras 615–45.
27. *Royal Commission on the Constitution, 1969–1973, Vol. II: Memorandum of Dissent* Cmnd 5460, London, HMSO (1973) pp. 104–5.
28. ibid. pp. 150–62.
29. *Minority Report*, op. cit. p. 147.
30. Diane Dawson *Revenue and Equilisation in Australia, Canada, West Germany and the U.S.A.* Research Paper 9, Commission on the Constitution, London, HMSO (1973) p. 83.
31. W. E. Oates *Fiscal Federalism* New York, Harcourt Brace Jovanovich (1972) p. 29.
32. G. McCrone *Scotland's Future: The Economics of Nationalism* Oxford, Blackwell (1969).
33. For example, Scottish gross domestic fixed capital formation is only some 9–10 per cent of the UK total. A 50 per cent increase in the level of capital investment in Scotland would increase total UK investment by only 5 per cent, equivalent to only 1 per cent of UK gross national product.

Index